EVOLUTION

EVOLUTION

A VISUAL RECORD

PHOTOGRAPHS BY ROBERT CLARK FOREWORD BY DAVID QUAMMEN
TEXT BY JOSEPH WALLACE

THESE VIEWS OF LIFE

David Quammen

It's no coincidence that Charles Darwin used a visual metaphor in what became his single most famed and quoted sentence. As published in the first edition of *On the Origin of Species*, 1859:

> There is grandeur in this view of life, with its several powers, having been originally breathed into a few forms or into one; and that, whilst this planet has gone cycling on according to the fixed law of gravity, from so simple a beginning endless forms most beautiful and most wonderful have been, and are being, evolved.

The "view" to which Darwin referred was of course not just an optical sighting but a scientific theory, a conceptual construct: evolution by natural selection. But it was a theory that had been erected, to a great degree, on the foundations of visual evidence. Darwin was a keen-eyed, attentive observer—one of his great strengths, crucial to the way he worked—and so was the co-discoverer of the theory, Alfred Russel Wallace (whose story you'll read in the fine essay by Joseph Wallace—no relation— that follows here). Both men, Darwin and Wallace, were natural historians who traveled the world with their eyes wide open. Their brains were ever eager to make sense of the facts, the peculiarities, the patterns, but their data were primarily visual. In more recent decades, by contrast, evolutionary biology has been much enhanced and deepened by abstract and less visual forms of evidence—notably, by the coding sequences recorded in DNA, now easily read off by fancy machines and comparatively analyzed with the aid of computers, in an enterprise called molecular phylogenetics—but the origins of this science lay in eyeball scrutiny and the deductions to which such eyeballing led. That was the first tool of that first great pair of evolutionary thinkers, and of many who followed them, from Ernst Haeckel to Ernst Mayr to Edward O. Wilson: *looking*. So it's fitting indeed that Robert Clark, a visual artist of surpassing skill and hard head, has devoted so much of his professional life to producing a visual record of evolution.

It's more than fitting, in fact. It's enormously valuable. It's illuminating in ways that words can't match. It's also very damn pleasurable, because it reminds us that, as Darwin said, "endless forms most beautiful and most wonderful have been, and are being, evolved." The beauty is generally accidental, secondary to evolutionary mechanisms and adaptive functions (except in cases of sexual selection, where females demand beauty of males in exchange for chances at reproduction), but those beauteous accidents are partly why we prefer a world full of evolutionary diversity to a dead planet or one impoverished by extinctions. Minerals can be pretty, but they're nothing like a birdwing butterfly or an elaborate orchid or an elegant coppery snake.

Robert Clark knows that and has known it for a long time. My first encounter with his work came when we were partnered by *National Geographic*, back in 2004, for an article that ran under the intentionally provocative title "Was Darwin Wrong?" The answer was: *No, not on evolution he wasn't*, and Rob's photographs supported that point eloquently. The evidence for evolution that Darwin presented in *The Origin*, as I explained in my text for that article, fell into four categories: biogeography, morphology, paleontology, and embryology. Biogeography is the study of which creatures live where on planet Earth, and why they live there but not elsewhere. Morphology is the study and comparison of their bodily shapes. Paleontology, working from the fossil record, tracks changes in those shapes through time. Embryology is the investigation of how those shapes develop before birth or hatching. Note that each one of those four categories of evidence, Darwin's evidence and the main forms of evidence for the following hundred years, is inherently visual.

Four years later, Rob Clark and I went out together for *National Geographic* again, this time on a story about Alfred Russel Wallace. While I spoke with experts, examined Wallace's notebooks, and ate dinner with Wallace's grandson, Rob photographed Wallace's beetle and butterfly specimens, preserved carefully in the warehouse of the Natural History Museum, in London. Wallace, a young man with driving curiosity but no money, had made his living as a professional collector of specimens, spending four years in the Amazon and then eight years in the Malay Archipelago, concentrating especially on beautiful creatures because they were the most saleable: birds of paradise, gaudy moths and butterflies, iridescent beetles. Amid the variations on beauty, Wallace saw nuances and patterns that led him toward understanding evolution—and Rob Clark, through his lenses, captured luminous images, glimpses for the rest of us, of what Wallace had seen.

Let me take you now on a short digression that points back toward the center. There is an extraordinary man named Geerat J. Vermeij, a Dutch-born paleontologist and evolutionary biologist, who stands as an exception that proves the rule. Dr. Vermeij, a Distinguished Professor at the University of California, Davis, has been blind since the age of three. Zero visual input. He studies the evolutionary history of marine mollusks. In place of visual data he has *feel*—his knowing, sensitive fingers explore the shapes of living and fossilized shells, perceiving the subtleties of ridge, crenulation, and curve. Vermeij has produced much-admired papers and books, offered influential ideas, won awards. His autobiography is titled *Privileged Hands*. He doesn't need vision to grasp how evolution works.

Most of the rest of us do. Darwin and Wallace did. Evolution is the shape of life's history, made manifest to the eye and the mind, and its particulars are here, in this gorgeous and telling book, for all lookers to see.

DARWIN, WALLACE, AND THE BIRTH OF A THEORY

Joseph Wallace

Once great scientific discoveries have been accepted as truth, it's hard to imagine what life was like before they came along. Did people ever truly think that the earth was flat, that salamanders were born out of fire, that birds of paradise were actually messengers from heaven?

Yes, they did, even if each of those ideas seems absurd to us now.

So it is with the theory of evolution. Today, most of us find it impossible to believe that every living species—from the tiniest patch of moss and beetle to the mightiest whale and redwood—was carefully placed on Earth and has existed unchanged ever since the Great Flood.

Yet only a century and a half ago most people *did* believe this, and not just those who gave it only a casual thought. Practically every leading Western thinker—scientists, philosophers, and, of course, clergy— accepted the biblical view of creation without question. And their faith in it was so unshakeable that anyone who might feel differently was swiftly condemned and ostracized.

ABOVE LEFT CHARLES DARWIN IN 1874, FIFTEEN YEARS AFTER PUBLISHING *THE ORIGIN OF SPECIES.* ANTICIPATING THE UPROAR FROM A SCIENTIFIC COMMUNITY STILL STEEPED IN THE BIBLICAL VIEW OF CREATION, HE HESITATED FOR TWO DECADES BEFORE MAKING HIS THEORY OF EVOLUTION PUBLIC. AS HE'D PREDICTED, HE WAS STILL EXPLAINING, DEFENDING, AND REFINING IT UNTIL THE END OF HIS LIFE.

ABOVE RIGHT ALFRED RUSSEL WALLACE MAY HAVE DRESSED UP FOR THIS PORTRAIT, BUT THESE WERE NOT THE SORT OF CLOTHES HE WORE FOR MUCH OF HIS LIFE AS A PROFESSIONAL WILDLIFE COLLECTOR IN SOUTH AMERICA, ASIA, AND OCEANIA. HIS SOLITARY ADVENTURES OVER THE COURSE OF MANY YEARS GAVE HIM THE EXPERIENCE, INSIGHT, AND TIME TO DEVELOP HIS OWN EVOLUTIONARY THEORY, WHICH SPURRED DARWIN TO PUBLISH *THE ORIGIN OF SPECIES.*

PAGE 10 THE SKELETONS OF AN ORANGUTAN AND A GIRAFFE DEMONSTRATE THE BREADTH OF MAMMALIAN EVOLUTION. ORANGUTANS WERE ONE OF THE SPECIES THAT DREW ALFRED RUSSEL WALLACE TO THE FORESTS OF SARAWAK ON THE ISLAND OF BORNEO, WHERE HE DEVELOPED HIS THEORY OF EVOLUTION. AND CHARLES DARWIN WROTE EXTENSIVELY ABOUT THE GIRAFFE, WHOSE LONG NECK — WELL ADAPTED TO TREETOP BROWSING — HE CONSIDERED A STRIKING ILLUSTRATION OF NATURAL SELECTION.

All of which makes it critical, when discussing and illustrating the wonders of evolution, that we remember the world in which the young Charles Darwin (1809–1882) and Alfred Russel Wallace (1823–1913) — the two men who independently figured out the most essential explanation for the extraordinary diversity and adaptability of life on Earth — lived. That we remember that they were brilliant for devising the theory, but also more than merely brilliant. They were courageous.

They had plenty to lose, but they went ahead anyway, one immediately and the other only after many years of worry and hesitation. And they did more than just propose a new way of looking at life on Earth: They demonstrated it so clearly and convincingly that it is now beyond the reach of logical debate.

Darwin was the first to develop a scientific theory to explain life's diversity, yet his ideas didn't arise from a void. In the years before the publication of *The Origin of Species* (1859), others had begun to take the risk of proposing that life on Earth had the capacity to change, to evolve — that it was (in the term of the time) mutable.

Those others included members of Charles Darwin's own family. Before Darwin was born, his grandfather, the prominent physician and philosopher Erasmus Darwin, had floated the idea that species could both "continue to improve" over time (a process then commonly called transmutation) and pass those improvements along to the next generation.

Having such a free thinker in the family undoubtedly helped set Charles on his own iconoclastic course. But Erasmus would go only so far, making sure to state his unequivocal belief that God's design was the instrument of the "improvements" he described.

Still, despite such doctrinaire insistence on a strict biblical interpretation of Earth's history, Charles was born into a time and place when testing the boundaries of scientific knowledge was more possible than it had been for centuries. Driving this surge of discussion and debate was the fact that the world was becoming more accessible than it had ever been. This new access was leading to discoveries that challenged not just the thinking about the origins of life on Earth, but many other long-held beliefs as well.

A main culprit in the shaking of the conventional wisdom was the development of great transoceanic sailing ships, which brought myriad European explorers and traders to corners of the earth that had until then been blank spaces on the map. The great colonial powers expanded their reach and jockeyed for position, using the world as their chessboard — and if you wanted to go along to study birds or bugs, you were welcome to.

As a consequence, what had been nearly legendary unvisited lands, remote islands, and oceans labeled "Here Be Monsters" were now demonstrably real, solid, earthly — and filled with unexpected marvels. Explorers

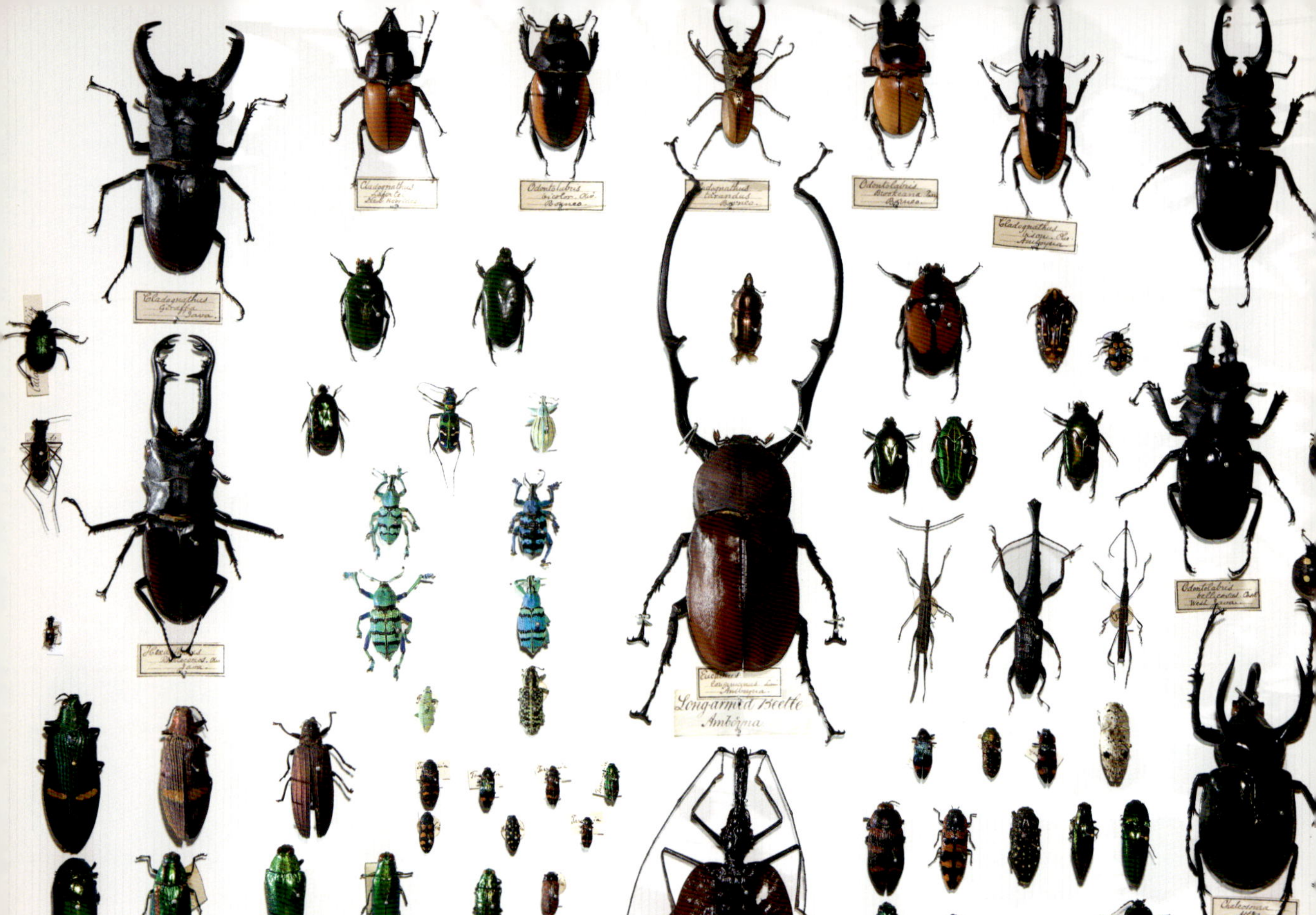

IF THERE IS ONE ENTHUSIASM THAT SEEMS TO HAVE BEEN SHARED BY MOST NINETEENTH-CENTURY COLLECTORS AND AMATEUR NATURALISTS, INCLUDING CHARLES DARWIN AND ALFRED RUSSEL WALLACE, IT WAS AN INTEREST IN BEETLES. ALONG WITH BEING COLORFUL, BEETLES WERE — AND ARE — EVERYWHERE, COMPRISING ABOUT 40 PERCENT OF ALL DESCRIBED INSECT SPECIES. IN SHOWING BOTH THE DIVERSITY AND THE CLEAR RELATIONSHIPS AMONG VARIOUS SPECIES, THIS BOX OF SPECIMENS FROM WALLACE'S OWN COLLECTION PROVIDES BEAUTIFUL EVIDENCE OF THE MAGNIFICENT DIVERSITY OF THE NATURAL WORLD BROUGHT ABOUT BY EVOLUTION.

found tropical forests filled with an almost incomprehensible number and variety of plants, birds, and beetles. The African plains were flooded with game in equally stunning profusion, while islands hosted bizarre flightless birds and other fantastical creatures (including those messengers from heaven, the birds of paradise, native to the island of New Guinea). Witnessing all this and more, explorers and scientists back home alike inevitably began to ponder what exactly it was they were seeing, and what it all might mean.

Among the first — and certainly among the most famous — of those to propose a theory of a non-static living earth was the French scientist and writer Jean-Baptiste de Lamarck (1744–1829). His books on plants and invertebrates made him a leading figure in scientific circles, but it was his 1809 *Philosophie Zoologique* that brought him renown.

Today, Lamarck remains famous mostly for what he got wrong: He was the man who actually believed that living creatures could (by means of substances known as "subtle fluids") dramatically change their shape during their lifetimes—and then pass these changes along. The best-remembered example of his thinking is the giraffe: A single individual could significantly lengthen its neck to reach leaves higher in a tree, and then bequeath this improvement to its young, all of whom would be born with long necks.

Still, by building from the idea that mutability could change a species forever, Lamarck's conclusions were a significant step forward from previous ones. And, though Darwin later went to great lengths to emphasize that he did not subscribe to Lamarckism, the earlier scientist's conclusions clearly sowed some seeds for Darwin's own.

At the same time that the concept of species' immutability was being questioned, so was the long-held belief that every species would continue to exist forever after having been placed on Earth by the Creator.

Again, it was the energetic spirit of the era that challenged this bedrock-solid belief. Both close to home and across the world, the earth itself was revealing unexpected mysteries as people avidly tilled new farmland, built countless roads and railroads, and mined coal and iron on a large scale. These unearthed treasures included the gigantic teeth and bones of elephant-like mammals (even in places like North America and Europe where no living elephants were known to exist) and bones that resembled those of sloths and bears but were enormously larger than ones belonging to any known living species

How to explain such astonishments? Many leading scientists refused to accept that the creatures that had left these remains had truly vanished. With enough searching, they would turn up somewhere on Earth. They had to.

But one of the effects of enthusiastic travel and exploration is that the blank spaces on the map shrink, and then disappear entirely. If you don't find the monsters in the places labeled "Here Be Monsters," it's time to redraw the map and adjust your expectations.

So, by the early decades of the nineteenth century, Darwin's formative years, even some previously skeptical experts had begun to accept that the history of life on Earth was different than they'd imagined—and than their religious upbringing had taught them to believe. It had to be faced: Not every species that had once walked the planet still did.

The secrets no longer hidden under the earth's surface stirred even greater scientific tumult. Studying the rock layers brought to light, geologists like Georges Cuvier (1769–1832) and Charles Lyell (1797–1875) were coming to another important conclusion: The earth's geologic record—if not its exact age—was revealed in successive, clearly distinguishable layers. And what's more, these layers indicated that the planet was far older than biblical teachings implied.

Then there were the bones and teeth that—unlike those of mammoths and cave bears—did not appear to closely resemble any creatures that still walked the earth. For decades, fossil hunters had been finding the remains of extinct animals that had been dozens of feet long with features reminiscent of modern lizards. But they were nothing like modern chameleons or skinks.

It was anatomist Richard Owen who in 1842 first coined the term *Dinosauria*, usually translated as "terrible lizards." (Though "fearfully great" or "scarily huge" lizards would be more accurate.) Unsurprisingly, the fact that something as spectacular as dinosaurs had once existed, yet by all appearances had entirely vanished, caused a sensation. (And, as movies like *Jurassic World* show, it still does.)

But what on Earth were they? And why had they disappeared, leaving only their fossils behind? Agreement—or even logical explanation—was hard to find, even among the era's leading minds.

For example, Lamarck—one of the first to propose the concept of evolution—refused to believe that species ever went extinct. Cuvier, on the other hand, was fine with extinction (he thought that catastrophic floods and other natural events periodically wiped species from the earth), but scoffed at any concept of evolution. New species would be placed where others had vanished and remain there forever . . . or at least until the next catastrophe, when the process would be repeated.

If there ever was a time, filled with discovery and uncertainty alike, when the scientific establishment and the public were ready for a revolutionary—and evidence-based—new theory of how life appeared on Earth and developed its abundance and diversity, it was in late 1831, as the HMS *Beagle* departed from England for South America.

Yet it would have been impossible to predict that the *Beagle* would be the birthplace of this theory. Much less that its developer would be the ship's on-board naturalist, a young man of twenty-two, whose at-times-indifferent studies at school had delayed (forever, as it turned out) his plans to be a doctor or a parson; who had no formal scientific training; and who'd been invited aboard as much to provide gentlemanly companionship to its captain, Robert FitzRoy, as to study the birds, bugs, and other beasts the voyage would inevitably encounter.

But this wasn't just any young amateur naturalist. This was Charles Darwin, who'd actually spent his seemingly aimless youth honing the irreplaceable skills that, in fact, made him the perfect candidate to change the way we see the world.

"I am a firm believer," Darwin said, "that without speculation there is no good and original observation."

It was a simple credo, yet it lay at the heart of all the discoveries he made on the *Beagle*'s long voyage, just as it had throughout his life. Look, but don't just look, *question*.

Though never a trained scientist, Darwin had been a dedicated amateur naturalist since childhood, "obsessed with beetles in a way that other boys are obsessed with marbles," as the writer Adam Gopnik puts it. Charles

CHARLES DARWIN ALWAYS LOVED BEETLES. "I WILL GIVE PROOF OF MY ZEAL: ONE DAY, ON TEARING OFF SOME OLD BARK, I SAW TWO RARE BEETLES, AND SEIZED ONE IN EACH HAND; THEN I SAW A THIRD AND NEW KIND, WHICH I COULD NOT BEAR TO LOSE, SO THAT I POPPED THE ONE WHICH I HELD IN MY RIGHT HAND INTO MY MOUTH," HE ONCE WROTE IN A LETTER. "ALAS! IT EJECTED SOME INTENSELY ACRID FLUID, WHICH BURNT MY TONGUE SO THAT I WAS FORCED TO SPIT THE BEETLE OUT, WHICH WAS LOST, AS WAS THE THIRD ONE." SO HE MUST HAVE BEEN PLEASED WHEN THIS INDIVIDUAL (FAMILY PASSALIDAE) LANDED ON THE *BEAGLE*, NOT ONLY PROVIDING HIM WITH A NEW SPECIMEN, BUT DEMONSTRATING FIRSTHAND ONE WAY THAT ANIMALS — INCLUDING SMALL INSECTS — CAN REACH AND THEN POPULATE ISLANDS THAT, LIKE THE GALÁPAGOS, MAY LIE HUNDREDS OF MILES OFFSHORE FROM THE NEAREST LANDMASS.

was happy spending hours alone in the woods near his home in Shrews-bury, England, because when he was surrounded by wildlife — and especially the insects that fascinated him — he never truly felt alone.

These local explorations must have helped him develop not only his powers of observation and ability to question what he was seeing, but also his patience, which was at least as important. As he would prove many times in years to come, he understood the value of thinking things through.

Still, nothing in the English countryside could have prepared him for the tropical forests he visited on the *Beagle*'s early stopover in Brazil. As they have for many explorers whose native climes are northern, the mind-boggling abundance and variety of life he found there nearly over-whelmed him.

"He was awed as usual on one such walk, and his eyes were unable to settle," Lyanda Lynn Haupt writes in her 2006 book *Pilgrim on the Great Bird Continent*. "Finally he reached for his pocket notebook, want-ing to capture something of this strange, soft elation. 'Twiners entwining twiners — tresses like hair — ' he wrote quickly, as if taking dictation '—beautiful Lepidoptera—' until he was barely within the realm of words, '—Silence—,' and finally exclaimed, quietly to himself and to the movement of this twining earth, '—Hosannah—.'"

But as the *Beagle*'s voyage went on, exposing Darwin to a range and variety of environments and the creatures that inhabited them in Argen-tina, Chile, and elsewhere, he moved past the nearly inarticulate sense of revelation he'd felt in Brazil. While never losing his sense of wonder, he marshaled the questioning spirit he'd honed during his childhood and engaged in the speculation that, by his credo, made for the only useful observation. For example, in many places along the route — including at thirteen thousand feet in the Andes Mountains — he found the fossils of seashells. His notes from the voyage are filled with questions about the geologic processes that caused the surface of the planet to rise and subside over great periods of time.

Even so, Darwin did not develop his theory during the *Beagle*'s five-year journey. Only when he returned home in 1836 did he have the time and comfort to start working out the implications of what he'd observed and collected in the rainforests and on the coasts of the great South American continent — and also on the volcanic islands, scattered in the Pacific Ocean six hundred miles from the Ecuadorean mainland, that were known as the Galápagos.

Back in England, he sought out some of his time's, and his nation's, leading scientists. These were experts with far more training than Darwin had ever received, and their reactions to what he'd brought back helped lay the groundwork for his future leaps of imagination.

First he showed his bird collections to the noted ornithologist John Gould. Gould pointed out that a group of small birds Darwin had collected—and, due to their remarkable variety in bill size and shape, had thought unrelated—were, in fact, closely allied. These were the famous "Darwin's finches," but at this point they were just minor players in the development of his theory of evolution, because Darwin had neglected to label which species came from which island.

Fortunately, Gould found a similarly curious—though less visually spectacular—example of the same phenomenon among the islands' mockingbirds. At least three different mockingbird species inhabited the Galápagos, each one living on a separate island with no overlap. Importantly, though their differences made Gould sure that they were all unique species, they still resembled each other—and certain species found on the South American mainland—closely enough to raise questions in Darwin's mind.

Countless others before him had observed different species that resembled each other. But it was Darwin who reached a new conclusion from this information: Given the essential mutability of species (which was then still a controversial topic in scientific circles) and the right conditions—geographic isolation, time—a single bird species could evolve into multiple new ones.

Today, in hindsight, this insight seems almost too obvious to be worth mentioning. At the time, however, it was an imaginative leap worthy of being dubbed "the first Darwinian revolution" by the great twentieth-century evolutionary biologist Ernst Mayr.

But if living things' capacity to evolve into new species—reflected everywhere he looked—came clear to Darwin soon after his return to England, the possible consequences of making his theory public were

TYPE. Brit. Mus. Reg. 55. 12. 19. 14.
E. magnirostris
...spiza strenua Gould
Chatham Id. Pres. by Zool. Societs.
Nᵒ TYPE. Brit. Mus. Reg 55. 12. 19. 20.
Cactornis scandens, Gld.
Ja.mes Island Ex. Coll. Darwin.
Vᵒ
Brit. Mus. Reg. 56. 3. 15. 4
Geospiza strenua
Pres by C. Darwin
Pc.
Brit. Mus. Reg. 56. 3. 15. 4

equally obvious. As of 1844, eight years after the end of the *Beagle*'s voyage, he was still struggling to express his thoughts out loud, even in private correspondence. "At last gleams of light have come," he wrote in a letter to a friend, the botanist and explorer Joseph Hooker, "& I am almost convinced (quite contrary to opinion I started with) that species are not (it is like confessing a murder) immutable."

It is like confessing a murder. Since (as David Quammen puts it in his book *The Reluctant Mr. Darwin*) "Truth be told, by 1844 he was more than 'almost' convinced" — and had been since the previous decade — Darwin's hesitation to own up to it was based not on uncertainty about the facts, but on the tempest he knew his theory would unleash.

Early on, he may have told himself that he didn't have enough information yet. The birds of the Galápagos proved beyond a doubt that transmutation — evolution — clearly took place, and was in fact the driving force behind the diversity of all life on Earth. But how did it work? What was the mechanism, the spur to change?

In 1838, again through an encounter with a leading thinker (though this time via the written word), Darwin came up with his answer. The inspiration was *An Essay on the Principle of Population* (1798) by Thomas Malthus, which proposed a doomful view of the future.

The basis of Malthus's argument was that populations of living creatures are capable of increasing geometrically, but food supplies can increase only arithmetically. This, Malthus believed, would sooner or later lead to apocalypse, as the human population (especially the population of the poor) exploded, far outstripping the ability to feed it.

Once they caught on with the public, Malthus's theories had notorious real-world consequences, including the draconian Poor Law in Britain, which forced husbands and wives on welfare to live separately in an effort to keep them from having "surplus" children. But they also led Darwin to ask a crucial question: If species were capable of multiplying geometrically, why didn't they? After all, it took only a cursory glance out the window to see that the world was not, in fact, being overrun by geometrically increasing animal populations.

In a breathtaking insight, Darwin combined his knowledge that species can and do evolve with the question of why their populations don't overrun the planet, and came up with the missing piece of his great theory: the concept of natural selection. This is the idea that survival for a given species in a given habitat is an ongoing struggle — one that keeps the population from decimating the food supply. Further, a species' mutability is both the key to its survival and the engine driving the evolution of new species.

Or, as Darwin himself put it in his *Autobiography*, "It at once struck me that under these circumstances favourable variations would tend to be preserved, and unfavourable ones to be destroyed. The result of this would be the formation of new species."

Darwin ends this passage with a casual "Here, then, I had at last got a theory by which to work." But from the start, he recognized how monumental his seemingly simple statement was.

As the writer Peter Bowler puts it in *Evolution: The History of an Idea*, at a time when many scientists still believed (in the face of all tangible evidence) that all species and all individuals within each species were immutable, Darwin defined species instead "as a population of unique individuals united only by the fact that they are potentially capable of breeding together. The species *is* the population, whatever the amount of variation in physical structure between the individuals concerned. . . . Variation is not a trivial disturbance of the ideal form but an essential character of the population and hence of the species."

Suddenly it all came clear: The fact that life forms can change, powered by natural selection, explained why (given enough time) new species could appear; why closely related — but not identical — species (like the Galápagos mockingbirds and finches) could coexist; and why species neither overran nor succumbed to environments that (as geologists had shown) themselves were always changing.

It's impossible to resist using Lamarck's famous example of the giraffe's neck to illustrate how Darwin's theory of natural selection worked. Like Lamarck, Darwin believed that a short-necked giraffe species could eventually become a long-necked one, but not because of "subtle fluids" acting on single individuals and their offspring.

Instead, the general tendency of species to change meant that sometimes giraffes would be born with longer necks, different configurations of teeth, stubbier legs, or different markings, just by chance. Then natural selection could act.

Most such random mutations would provide no advantage to the individual. Not being "selected," they would have no impact on the overall population.

But say that food became less abundant due to a drought or an infestation of leaf-eating insects. Suddenly those individuals with a longer neck would have an advantage, allowing them to reach leaves that were inaccessible to the others.

By living for more years and in better health, these longer-necked individuals would have a greater chance to breed successfully. The result of that breeding (because, as we now understand, the longer-necked animals would pass along the genetic mutation responsible for that trait) was an increasing number of long-necked giraffes, better suited to weathering the demands of their environment.

Over the course of time, the early giraffe-like creatures living on the great plains of Africa — where trees are scarce and those that exist often keep their leaves as far from the ground as possible — evolved into the long-necked species we know today. Scientists now recognize at least nine different giraffe subspecies (with differences in size, coat pattern, and other features), and believe that some might actually warrant full-species status.

Conversely, though it evolved from a common ancestor (possibly *Climacoceras*, which lived in what is now Kenya about fifteen million years ago), the giraffe's closest living relative — the okapi — did not develop a longer neck. Why not? We can never know for sure, but a logical supposition would be that in the dense, dim highland forests the okapi calls home, where its preferred diet — leaves and buds of shrubs — is abundant close to the ground, a long neck would actually be a disadvantage. So other traits (such as strong night vision) were favored as this species evolved.

The same process holds true throughout the broader story of evolution. For example, the world's eight surviving bear species clearly evolved from a common ancestor (most likely *Ursavus*, which lived roughly thirty million years ago), and most are still closely related. Yet the polar bear, inhabiting a frigid, snowy landscape frequently devoid of vegetation but filled with animal protein, is huge, blubbery, nearly entirely carnivorous, and white in color. Its cousin, the black bear, however, feasts on berries and other plant material along with meat, is far smaller, and is almost always garbed in black, brown, or cinnamon, earth-tone colors suited to its environment.

And it's not just mammals, of course. Throughout the immense web of life on Earth (scientists estimate that there are approximately ten thousand species of birds and at least a million insect species alive today), the story is the same: a series of adaptations has been selected for by their success in meeting the challenges and requirements of a species' environment, leading to the seemingly (though in reality ever-shifting) "ideal"

IN 1855, NEARLY TWENTY YEARS AFTER HIS RETURN ON THE *BEAGLE* AND NEARLY AS MANY SINCE HE'D FIRST DEVELOPED HIS THEORY OF EVOLUTION, DARWIN TURNED HIS ATTENTION TO THE BREEDING OF PIGEONS. HE ENJOYED THE CHALLENGE, BUT HE ALSO HAD ANOTHER GOAL: THROUGH WHAT HE CALLED "ARTIFICIAL" OR "DOMESTIC" SELECTION, BREEDERS CAN QUICKLY REPLICATE THE KINDS OF DRAMATIC PHYSICAL CHANGES THAT TAKE EONS IN NATURE. DARWIN CREATED EVIDENCE FOR HIS REVOLUTIONARY THEORY IN HIS OWN PIGEON LOFT. LABELED IN HIS HAND, THIS SKELETON FROM HIS COLLECTION WAS FROM THE BREED THEN CALLED DRAGON (NOW KNOWN AS DRAGOON), ORIGINALLY BRED AS HOMING OR CARRIER PIGEONS, WHICH DARWIN HIMSELF RAISED.

----

match we see today between each species and its habitat. Whether it's hummingbirds evolving an extraordinary variety of bill shapes and lengths to enable them to drink nectar from an array of wildly different flowers, or tasty viceroy butterflies mimicking the vivid orange and black of poisonous monarchs, or Venus flytraps occupying a niche few other plants inhabit by evolving the ability to capture live prey, examples are everywhere.

By 1838, two years after the *Beagle*'s return, Charles Darwin had the how and the why of his theory in his grasp. What did he do with it next?

Nothing, pretty much. For two decades. Two whole decades during which, outside of letters to friends and a few others, he stayed silent about his great imaginative leap.

IT WOULD BE EASY TO ASSUME THAT CHARLES DARWIN AND ALFRED RUSSEL WALLACE —
THE TWO VISIONARIES WHO INDEPENDENTLY CAME UP WITH THE THEORY OF
EVOLUTION — WERE RIVALS, BUT THAT WOULD BE WRONG. WALLACE HIMSELF TOLD
THE CINCINNATI ENQUIRER IN 1887, AFTER DARWIN'S DEATH, "WE WERE GOOD
FRIENDS. . . . I HAVE MORE THAN ONE HUNDRED LETTERS FROM HIM." IN THEIR LONG
CORRESPONDENCE (INCLUDING THIS LETTER) THEY COVERED TOPICS FROM TROPICAL
WILDLIFE TO DOMESTIC POULTRY TO THE TRIALS AND TRIBULATIONS OF PROMOTING
A CONTROVERSIAL NEW THEORY. WHEN THE *ENQUIRER* ASKED IF HE STILL BELIEVED
IN THE THEORY OF EVOLUTION DECADES AFTER ITS PUBLICATION, WALLACE ADDED,
"MORE THAN EVER. THE LONGER A MAN LIVES AND STUDIES, THE MORE HE IS
CONVINCED OF THE TRUTH."

Not that he sat still during those twenty years. Far from it. He married Emma Wedgwood, his most trusted friend and reader, and together they had ten children. Together they bore the death of two of the children, including ten-year-old Annie (Darwin's acknowledged favorite), whose death devastated them. He battled the chronic intestinal and other illnesses that plagued him through his life.

He wrote books, but not *the* book.

During that time, he also plunged into intense study of artificial selection, actively breeding show pigeons to reinforce certain traits and learning about the creation of dog breeds. He spent year upon year studying and writing about barnacles, uncovering their evolutionary adaptations.

But it's impossible to look at those years without seeing caution shading into fear — and even paralysis — in Darwin's unwillingness to publish what he believed in so deeply. When he said that admitting to the concept of mutability of species "is like confessing a murder," he meant it.

Historians and students of science have long debated whether Darwin would *ever* have put his theory down on paper if someone hadn't arrived who inspired a deeper fear in him than even publishing it seemed to.

That someone, of course, was Alfred Russel Wallace, and what he did was come up with the theory of evolution on his own. Beginning, as Darwin had, with the concept of species mutability, and then later adding the idea of natural selection as its driving force, Wallace never developed his theory as thoroughly as Darwin did. But he was willing to publish his ideas as soon as he came up with them. No years of hesitation for him.

Wallace couldn't afford to hesitate. Unlike Darwin, who had possessed enough of a family fortune to live a life of exploration and then thinking and writing (and thinking about writing), Wallace had grown up in comparative poverty. While Darwin had joined the *Beagle* as a naturalist without any specific portfolio, Wallace traveled as a commercial collector, paying his way by collecting and subsequently selling as many specimens as he could.

And, while Darwin returned from the *Beagle*'s voyage and never left the United Kingdom again, Wallace didn't have the option to sit for years and ponder. After his initial trip to South America—which began in 1848 and culminated in 1852 with the sinking of his ship home and loss of his specimens and most of his notes—he headed out again in 1854. This time his destination was the Malay Archipelago (now Malaysia, Singapore, and Indonesia), where he spent eight years exploring widely, collecting avidly, and coming up with his own theory of evolution.

The results of Wallace's day-to-day activities may have helped inspire his theory. While a naturalist like Darwin collected just enough specimens to study—a couple of each species at most—Wallace aimed to collect as many of a wanted species as he could. Looking over the results of his efforts, he couldn't fail to notice variations in coloration, size, and other characteristics within even a single population of a species.

PAGES 28–29 WALLACE MARVELED AT THE "RICH BLUES AND CRIMSONS, THE DELICATE GREENS, YELLOWS, AND PURPLES, THE VELVETY BLACK AND PURE WHITE" OF THE PITTAS (FAMILY PITTIDAE, A GROUP OF MOSTLY TERRESTRIAL BIRDS CONFINED TO THE OLD WORLD). HE COLLECTED THESE SPECIMENS ON BORNEO AND SUMATRA, ISLANDS THAT LIE WEST OF WHAT IS STILL CALLED THE "WALLACE LINE." WALLACE HIMSELF WAS STRUCK BY THE FACT THAT MAINLY ASIAN FAUNA ARE FOUND TO THE WEST OF THE LINE AND AUSTRALIAN TO THE EAST, MAKING THE REGION A PRIME EXHIBIT OF HOW GEOGRAPHIC ISOLATION CAN LEAD TO THE EVOLUTION OF NEW SPECIES.

His brilliance was not in noting this—earlier collectors undoubtedly had as well—but in asking what it meant. Individuals within a species clearly weren't identical, and species on different islands or opposite riverbanks appeared similar but were not the same. Why, and why not?

Regardless of exactly how he got there, in 1855 Wallace—waiting out the monsoon season in Sarawak on the island of Borneo—wrote "On the Law which has Regulated the Introduction of New Species." It was published later the same year in the *Annals and Magazine of Natural History*.

Wallace's theories left crucial questions unanswered. But when Darwin read the young collector's italicized statement that *"every species has come into existence coincident both in space and time with a pre-existing closely allied species,"* he knew the details didn't matter. If he weren't to forfeit all credit for the theory he'd nurtured for so long, Darwin knew he'd better go public at last.

Charles Darwin's *On the Origin of Species by Means of Natural Selection or the Preservation of Favoured Races in the Struggle for Life*, first published in 1859, is now known by the much simpler title *The Origin of Species*. But the full title is worth celebrating because it shows that, having finally been goaded into action, Darwin opened the floodgates wide, without any equivocation.

For example, from the book's opening pages, the passionate statement of his certainty that natural selection is the driving force of evolution: "As many more individuals of each species are born than can possibly survive; and as, consequently, there is a frequently recurring struggle for existence, it follows that any being, if it vary however slightly in any manner profitable to itself, under the complex and sometimes varying conditions of life, will have a better chance of surviving, and thus be *naturally selected*. From the strong principle of inheritance, any selected variety will tend to propagate its new and modified form."

And, near the end of the book, this gorgeous proclamation of the glories of evolution as a whole: "There is grandeur in this view of life, with its several powers, having been originally breathed into a few forms or into one; and that, whilst this planet has gone cycling on according to the fixed law of gravity, from so simple a beginning endless forms most beautiful and most wonderful have been, and are being, evolved."

In for a penny, in for a pound. In the end, Darwin proved himself ready and willing to face the controversy he had known would erupt at his conclusion that nature itself—and not a deity—was responsible for the marvelous abundance and diversity of life on Earth. Crucially for an argument that needed to be convincing to both scientists and the public, he overcame both his fears and the intense time pressure spurred by Wallace's unexpected appearance and wrote a wonderful book.

Even to today's reader, *The Origin of Species* is clearly argued, vividly described, and filled with the joy of discovery, a welcoming of debate, and the adrenaline rush that characterizes intellectual leaps into the void. It also shows that Darwin's theories had developed and deepened in the years since his return to England.

For example, the book opens with a detailed exploration of "Variation Under Domestication," incorporating Darwin's own work with pigeons and his studies of dog breeding. By doing so, it provides an approachable entry—everyone was familiar with dog breeds and domesticated birds—to the more daring extensions of his statement on what the mutability of wild species meant about life on Earth.

As Darwin thought, hoped, and feared it would be, *The Origin* was a sensation from the day it was published, and a source of often enraged contention from that day forward. (As, in a few circles, it remains even today.) He wrote several updated editions during his lifetime, bolstering and fine-tuning his arguments.

But even in the face of the most withering criticism from religious leaders and scientists who refused to give up their creation-based views, Charles Darwin never relinquished his own belief—no, his *certainty*—that life on Earth was indeed a miracle, but a miracle of nature. And regardless of the controversies, then and now, the essential beauty and clarity of his revolutionary theory remain undimmed.

ANCIENT HISTORY

It's sometimes tempting to think of evolution as an elegant, purposeful process, endlessly striving toward some perfect match between a species' form and its environment. But of course that's not true: It is actually a vast, grinding engine of random mutation and change, filled with dead-ends, wrong turns, and irrelevancies, along with the seemingly ideal forms we see all around us. (And they, of course, are also fuel for the same engine, always undergoing change as well.)

The process by which evolution works—which Charles Darwin and Alfred Russel Wallace first figured out, and Darwin described in impossible-to-deny detail—explains the vast abundance of life on Earth. (How much abundance? Just among the animals, scientists estimate that roughly five thousand species of mammals, ten thousand birds, and between one and thirty million insects currently exist. *Roughly* being the operative word.)

But nearly every other question about evolution's history and the relationships among species remains open to debate and recalibration. (The fossil record—which seems to be bursting with evidence of everything from ancient ferns to dinosaurs—is, in truth, extremely spotty.) Evolution and our understanding of it are both perpetual-motion machines. That's what makes studying it so challenging and fulfilling.

PAGE 32 For decades, it seemed that the sole known ancient bird was the famous *Archaeopteryx*. In fact, there were always others, but only in recent years have paleontologists added greatly to the collection, and just as greatly to our knowledge of the evolution of birds. This crow-sized *Confuciusornis* lived in what is now China during the Early Cretaceous Period (125 to 120 million years ago). Since the first *Confuciusornis* specimen was found in the 1990s, paleontologists excavating the Yixian and Jiufotang formations in Liaoning have uncovered hundreds of complete *Confuciusornis* skeletons comprising at least four different species.

PAGES 34–35 Lichens, easy to ignore, tell a fascinating evolutionary story whose origins and path remain unclear. Each lichen species is a composite, made up of a symbiotic relationship between a particular fungus and an alga or cyanobacterium (or sometimes both). Even their classification has been long debated, although currently lichens are classified by their fungal component. Despite a spotty fossil record, evidence suggests that the earliest lichens appeared at least four hundred million years ago, and that their characteristic symbiosis has evolved many times independently at different times and in different places. Lichens come in many shapes, forms, sizes, and colors, including the bright red of this modern specimen photographed in northern Canada.

ABOVE AND OPPOSITE Before the eighteenth century, to a large degree humans lived and worked entirely on our planet's surface, hunting and gathering and tending small plots of land. Not until the rise of a more industrial age, with its massive expansion of mining (for coal, steel, and other products), farming, and excavation for building, did it become possible to read the history of Earth — and the organisms that had lived there — in the remains left behind in rock strata. This, in part, led to the development of the theory of evolution.

PAGES 38–39 Species evolve to survive gradual change in their environment — but what happens when conditions alter rapidly? The Joshua tree *(Yucca brevifolia)*, whose range is already limited mostly to portions of California's Mojave Desert, may provide an answer to this question. Due to climate change and other factors, it may become extinct in its main protected area (Joshua Tree State Park) by the end of this century. An ancient tree, it is unlikely to be able to spread to new habitats successfully: In the past, the dispersal of Joshua tree seeds appears to have been aided by the Shasta ground sloth *(Nothrotheriops shastensis),* which itself went extinct thirteen thousand years ago.

The fossil record is far from complete. It's impossible to know how many species have vanished without a trace, or how many fossils remain to be discovered. In recent decades, some superb specimens have been uncovered, such as those of this *Vancleavea campi*, whose teeth are still associated with its jaws, and bony scales with its skeletal remains. Such finds have allowed scientists to build their knowledge about the planet's extinct creatures, and as a result its evolutionary history. *V. campi* demonstrates both the treasure-hunt quality of paleontology and the challenges that result from an incomplete fossil record: Found only in a few locations in western North America, it is an unusual reptile (not a dinosaur) whose classification remains under debate to this day.

OPPOSITE In many fossil sites, the specimens are disordered, fragmentary, and incomplete. Occasionally, though — such as in the Santana Formation in Brazil and Green River Formation in Wyoming, Colorado, and Utah) — scientists find a gold mine: deposits containing fossils in such perfect condition that even the most fragile structures are preserved. This is a kind of seed fern called a glossopterid, and such pristine specimens have played a crucial role in our understanding of our planet's history. The distribution of glossopterids across now widely scattered continents helped scientists realize that Earth's landmasses were once joined in a single supercontinent. This led to a deeper understanding of plate tectonics and continental drift.

PAGE 44 The earliest fossils of palm fronds date back about ninety million years, to the Cretaceous Period. This places them in the midst of what appears to have been a great expansion in the number and variety of angiosperms (flowering plants) at the time. Paleontologists are able to learn in depth about the evolutionary history of angiosperms only in those rare cases when the plants were captured in sedimentary deposits or amber. The identification of this *Sabalites* palm frond preserved in sedimentary rock from the Green River Formation of Wyoming is not definitive, but it probably dates to the Early Eocene (56 to 47.8 million years ago).

PAGE 45 This exquisite *Waptia fieldensis* (which resembled modern shrimp, though its actual taxonomy remains unclear) comes from the Burgess Shale fossil beds of British Columbia. By combining great age (the Cambrian Period, 505 million years ago) with extraordinary preservation of features, these beds allow us to understand many details of the lives of the animals whose remains are preserved there. The Cambrian saw the first explosion of diversity of life on Earth, after around four billion years in which nearly all organisms had but a single cell. Within ten to twenty million years (a geologic eyeblink), the first ancestors of most species now alive on Earth — from algae to all backboned animals — appeared.

PAGES 46–47 In Charles Darwin's time, widespread excavation for mines, farms, and roads revealed something crucial: that neither the kinds of organisms that inhabit the earth, nor even the planet itself, remains static. One spectacular proof of this was provided by the discovery of the Western Interior Seaway, a vast, shallow ocean that in the Cretaceous Period ranged from what is now eastern Mexico all the way to northern Canada. How do paleontologists know that this area, mostly now far removed from any marine environment, had once been a sea? Because they found fossils belonging to the huge marine reptiles called the plesiosaurs and mosasaurs there, along with gigantic sharks, other fish, and crinoids (echinoderm relatives of sea urchins and sea stars) such as this beautiful specimen of *Uintacrinus socialis* from Kansas.

26 M

 Modern birders would immediately identify this specimen as a swift — but it's a swift (called either *Aegialornis szarskii* or *Scaniacypselus szarskii)* that lived during the Eocene Epoch, nearly fifty million years ago, in what is now Messel, Germany. Like many other spectacular fossils, it was found in the Messel Pit Fossil Site, an abandoned oil-shale quarry that was almost converted into a landfill before being saved for science; it is now a World Heritage Site. The bird's entombment in oil shale (formed by the gradual deposition of mud and vegetation in a wetland environment) allowed for its pristine condition.

PAGES 50–51 Charles Darwin lived in an era when new fossil discoveries were flooding into museums and universities. This bird fossil (an ancestor of modern mousebirds) was found in the Messel Pit. Insects, reptiles, and mammals have also been found in the pit, in such pristine condition that hair, feathers, stomach contacts, and even color have been preserved. Together, these specimens provide an unparalleled view of life during the Eocene.

PAGES 52–53 "The combination of good sedimentary conditions and the fact that animals, including hominids, like to be near a source of water," the great paleoanthropologist Richard Leakey says, helps explain why the remains of human ancestors — and many other creatures — are so often found near the shores of lakes. These beautiful human footprints, about 120 thousand years old, were discovered south of Lake Natron, Tanzania.

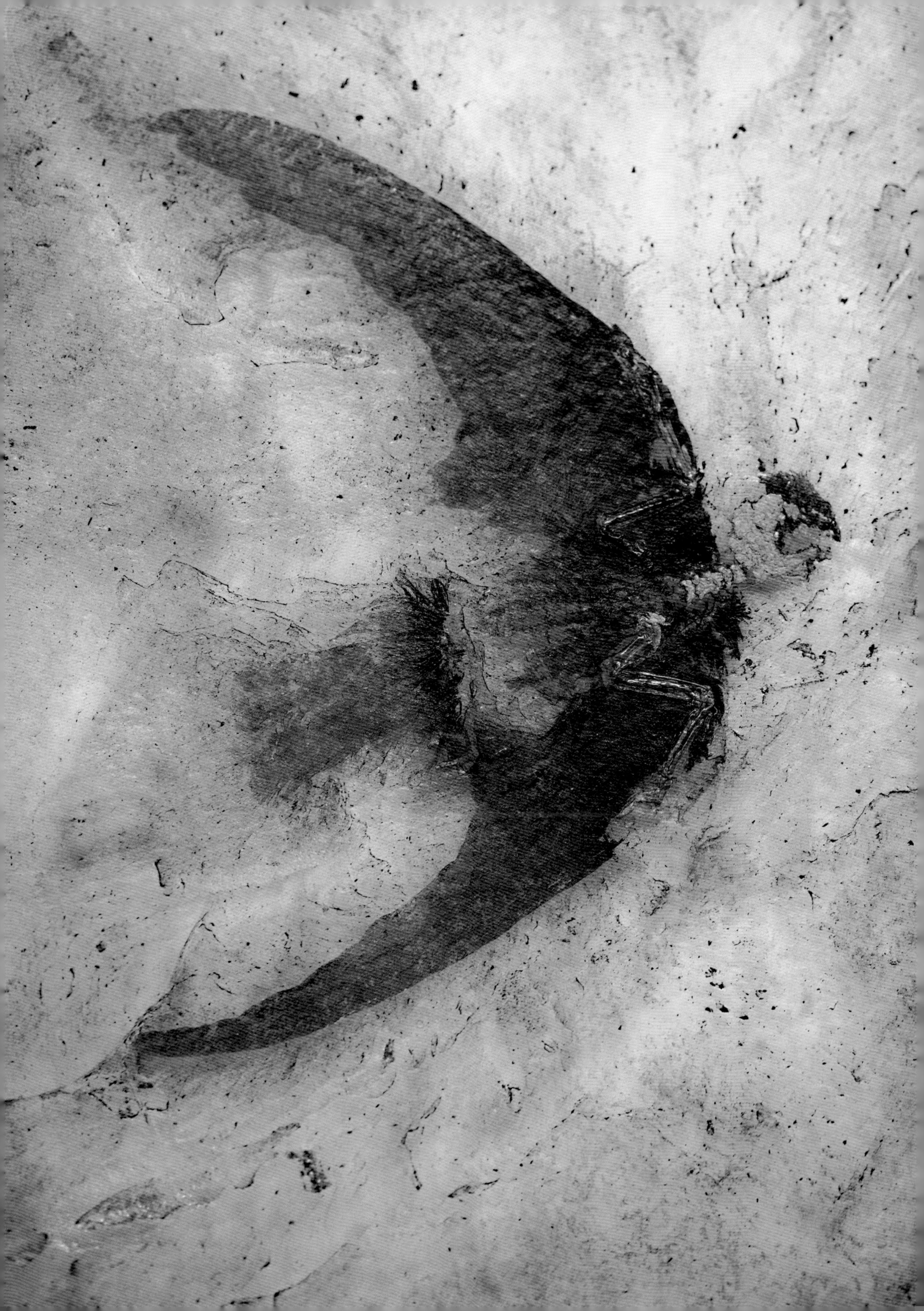

BIRDS
SELECTION, NATURAL AND ARTIFICIAL

The ability to fly has evolved repeatedly during the history of life on Earth: in countless insects; in one of the most widespread groups of mammals (the bats); and, of course, peerlessly, among the birds. (It is possible that amphibians will eventually join them as well; some amphibian species can glide long distances with the aid of skin flaps between their legs and torsos, much as the ancestors of bats likely did.)

Flight, having evolved so often and independently, is a good illustration of a pair of features of evolution that scientists have identified since Darwin's time. The first is convergent evolution, in which groups that are not closely related (such as mammals and birds) have evolved similar physical characteristics and habits and fill similar ecological niches. For example, hummingbirds and certain bats have both developed the ability to hover before flowers, drinking nectar.

Birds also illustrate the phenomenon known as parallel evolution, in which more closely related but geographically remote species evolve to resemble each other and exploit equivalent ecological niches. For example, rheas in South America, ostriches in Africa, and cassowaries in Oceania evolved separately to become huge, mostly vegetarian, and flightless.

OPPOSITE Not so long ago, it was a controversial theory, but now it's widely accepted: Birds aren't just dinosaur-like; they are in fact living dinosaurs. That's true of everything from sparrows to eagles to Darwin's finches — but it's rarely more obvious than when looking at a southern cassowary *(Casuarius casuarius),* the flightless bird native to Australia and New Guinea that at five feet tall and over one hundred pounds is one of the largest and heaviest birds on Earth. The steely gaze, the leathery crest, and the powerful beak all call the ancient dinosaurs to mind, as do the cassowary's fearsome three-toed feet, which (much as *Velociraptor* and other birdlike dinosaurs must have done) it can employ as weapons.

PAGE 58 This gyrfalcon *(Falco rusticolus)* specimen is just one of many prepared by the English ornithologist and taxidermist John Hancock during Darwin's time. Advances in taxidermy allowed Darwin to study birds he could never see in the wild, while the publication of his work also inspired the next generation of taxidermists. During the years he spent mulling over his theory, Darwin corresponded with many noted scientists, including John Hancock and his brother, Albany, also a naturalist.

PAGE 59 Birds, of course, are not the only animals that possess true flight. In a good example of convergent evolution, bats fill many of the niches at night that birds do during the day. While the evolution of birds has been a focus of scientific study since the discovery of the famous *Archaeopteryx* in 1861, bat evolution is less well documented. From what scientists understand, bats most likely came from an ancestor that used a membrane attaching its front legs to its body to glide from place to place, much as today's flying squirrels do. This gradually evolved into the active flight that has proven to be so successful today. About 20 percent of all mammal species on Earth are bats.

PAGE 54 AND ABOVE Since mutations are selected for the survival advantages they provide to a species, it's no surprise that the ability to fly has evolved repeatedly during the history of life on Earth, and in very different groups of animals. This phenomenon is known as convergent evolution. In ancient times, the reptiles known as pterosaurs — including such giants as *Quetzalcoatlus,* whose wingspan may have exceeded fifty feet — acquired active flight, a skill that persists in nearly all birds and bats, as well as a myriad of insects. The elegant structure of the wing of the South American cock-of-the-rock (genus *Rupicola*) is above; an anhinga wing (genus *Anhinga*) appears on page 54.

PAGES 60–61 "The sight of a feather in a peacock's tail, whenever I gaze at it, makes me sick!" That outburst was written by Charles Darwin in a letter to the botanist Asa Gray in 1860, just a year after the publication of *The Origin of Species*. Darwin wasn't railing against the beauty of the peacock and its stunning display. His problem was more basic: Since he believed that natural selection was driven by the challenges of finding food and avoiding being eaten, at first he simply could not understand what possible selective advantage a huge, cumbersome tail could provide to the bird. In fact, it should be a hindrance if a predator came by.

LEFT Darwin — and Alfred Russel Wallace, who was personally familiar with the Asian haunts of peacocks and their game bird relatives — came to believe that certain of the bird's features (including its size, elaborate coloration, and spectacular display) provided another kind of selective advantage: They helped attract mates. Recent research, however, suggests that the tail's huge size and prominent "eyespots" may also in fact intimidate predators, providing a survival advantage as well.

RIGHT The ocellated turkey, native to Mexico, Belize, and Guatemala, has only one close relative: the familiar wild turkey, whose domesticated breeds are the centerpiece of Thanksgiving meals in the U.S. The two species, the only members of the genus *Meleagris,* are both huge, relatively slow-moving, and very tasty to humans. As a result, the wild turkey came close to extinction a century ago, losing 98 percent of its original population before intensive reintroduction efforts rescued it. Its ocellated cousin (named for the ocelli, or eyes, that the spots on its tail resemble) has been less fortunate: It has vanished outside of a few protected areas, and is now considered a threatened species.

OPPOSITE Even using the most advanced analysis, scientists are sometimes still puzzled by the lineage of familiar species. One example is the lanner falcon *(Falco biarmicus),* native to Europe and Africa. Lanner falcons are usually placed with three other species in a small group known as hierofalcons. Though this group evolved around two million years ago, every living hierofalcon seems to be descended from the same common ancestor that lived only about 130,000–115,000 years ago. This seems to indicate an unexplained "bottleneck" in which the ancestor species nearly went extinct, but no one knows for sure.

RIGHT The evolution of the eye makes for one of science's most fascinating stories. Complex eyes appear in so many different species' lineage that scientists estimate that the organ has evolved independently as many as one hundred separate times. Recent research has also pinpointed how owls see so well in the dark: Their brains are specially designed to interpret the large amounts of information passed to them through the bird's outsize retinas, especially in low-light situations. This transfer and interpretation of visual information appears to be more crucial than either enhanced light-gathering capacity or sensitivity in allowing owls to be the exceptional nocturnal hunters most have evolved to be.

PAGES 66–67 If you go on an African safari, you might spot a four-foot-tall bird stalking by, its head and beak resembling an eagle's but its long legs looking more like a stork's. This is the secretarybird *(Sagittarius serpentarius),* a ferocious raptor related to kites and vultures. The bird prizes snakes — including venomous ones whose bite would quickly kill it. When it leaps on its prey, seeking to snap the snake's spine, only the bird's heavily scaled feet and legs are exposed. The more vulnerable parts of its body remain safely out of fangs' reach.

PAGES 68–69 Every chicken — regardless of its size, shape, plumage, or comb — is descended from the red junglefowl *(Gallus gallus),* which lives with three related species in the forests of India, Sri Lanka, Southeast Asia, and Indonesia. (And the chicken's yellow skin appears to have come from crossbreeding with the gray junglefowl, *G. sonneratii,* augmented by dietary additives contributed by farmers.) Humans first tamed junglefowl — possibly separately in Southeast Asia and southern India — at least seven thousand years ago. Since then, we've developed hundreds of distinct breeds, and poultry has become the fastest growing livestock group around the world.

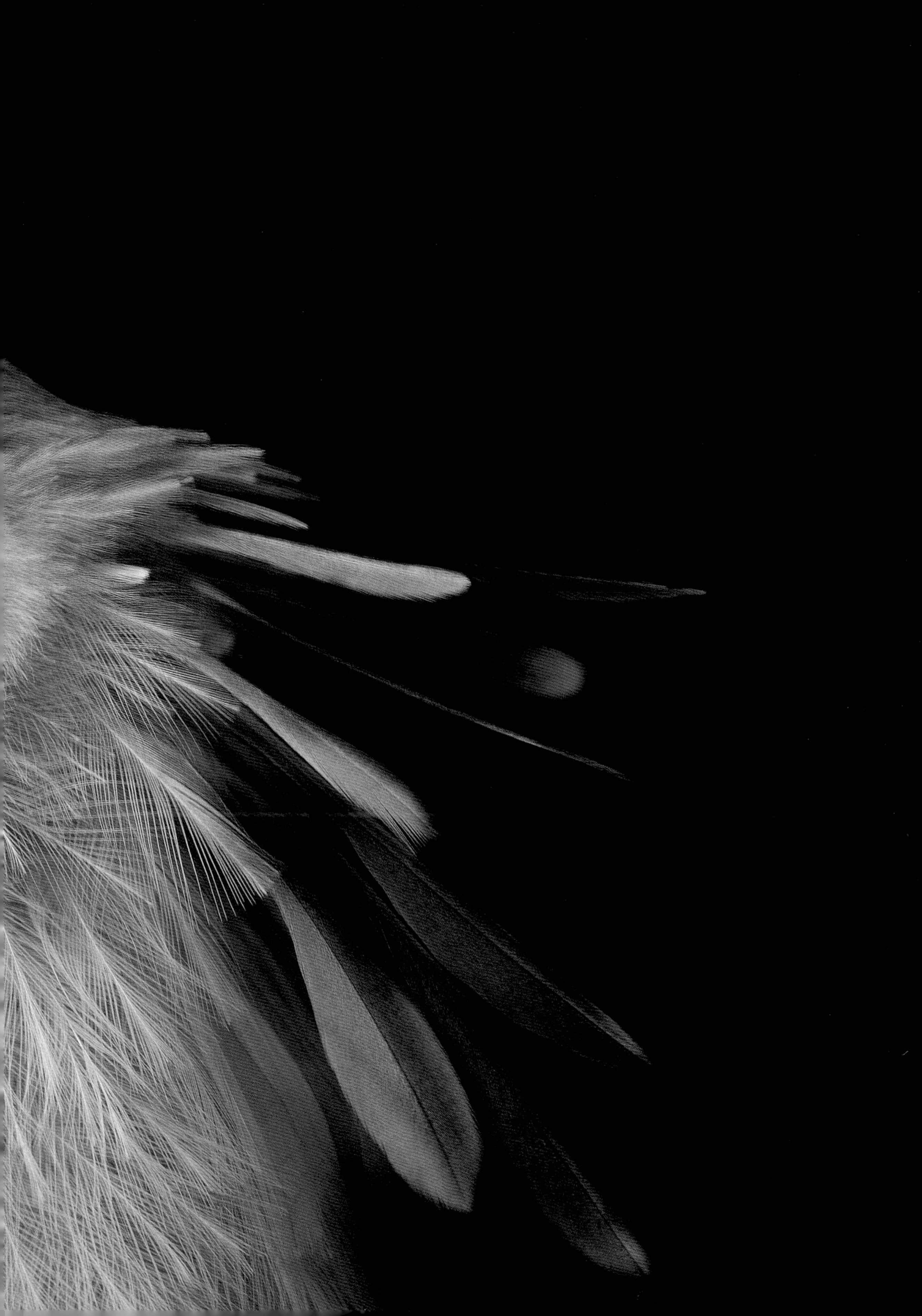

BELOW It would seem impossible for any bird to slam
its head into the trunk of a living tree up to twelve
thousand times daily, day after day, without sustain-
ing skeletal or brain damage. Woodpeckers are able
to do it because their brains are surrounded by a thick
layer of bone filled with microscopic bone fragments,
which form a dense, fibrous mesh that acts as a shock
absorber. In addition, the birds' hyoid apparatus (which
in humans is a modest bone-and-cartilage structure
for attachment of the tongue and some throat muscles)
is extremely large, wrapping around the skull to
provide more cushioning. Along with these and other
skeletal features, the birds instinctively vary the angle
of strike, minimizing repetitive-motion injuries.
Together, such adaptations have allowed woodpeckers
to thrive nearly worldwide in environments ranging
from boreal forests to the tropics.

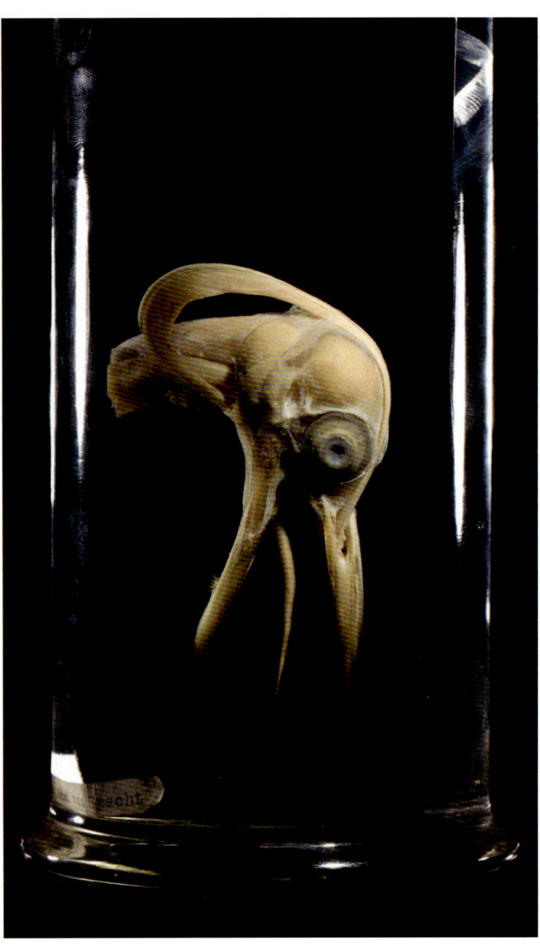

OPPOSITE As Charles Darwin discovered while
visiting the Galápagos and other archipelagos during
the Beagle's voyage, islands are home to some of the
earth's most bizarre — and most vulnerable — species.
Few islands demonstrate this more starkly than New
Zealand, whose avifauna once included nine species
of moa, gigantic flightless birds (lacking even the
vestigial wings of ostriches and emus) that could
reach twelve feet in height and weigh five hundred
pounds. Sadly, once the islands were occupied by
Polynesian explorers in the thirteenth century, moas
(with no defense against human hunters) disappeared
with astonishing speed. Every species was extinct
within two hundred years, leaving only a few skeletons,
such as this, behind.

PAGES 72–73 A glory of evolution, birds of paradise
are so spectacularly gaudy that early collectors
supposedly believed they were messengers from
heaven. Today, we know that all forty-odd Paradisaeidae
species evolved on what we now call New Guinea,
in northern Australia, and on nearby islands. Most
members of the family (including this female and male
lesser bird of paradise, Paradisaea minor) also developed
nearly unrivaled courtship behavior, in which the male
executes complicated dances, shows off his shining
feathers and shimmering plumes, and engages in
raucous vocal displays. Meanwhile, the drabber female
hops around, deciding if he's right for her.

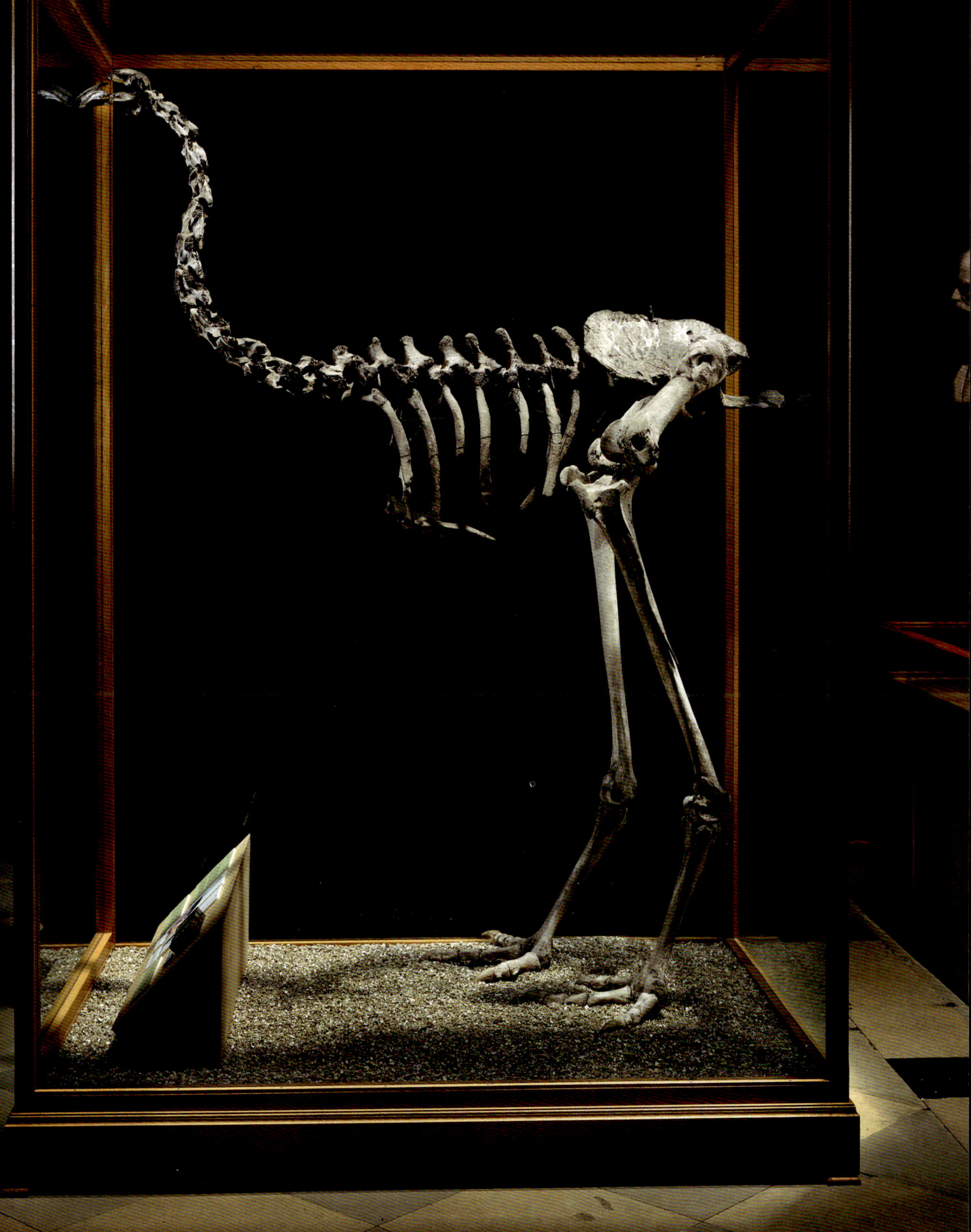

  This eye belongs to *Phasianus colchicus,* better known as the ring-necked pheasant. Native to Asia, the species has been widely introduced on other continents as a game bird. Dating back to at least the Oligocene Epoch (about thirty million years ago), pheasants are part of the same order, Galliformes, that includes junglefowl (and chickens), peafowl, quail, grouse, and turkeys. One characteristic of the group: They are tasty to humans and other predators, and thus many species are rare or declining in the wild.

OPPOSITE BELOW Like dozens of other bird species ranging from macaws to falcons to antbirds, the diamond dove *(Geopelia cuneata),* native to Australia, features colorful bare skin around the eyes. The scientific explanations for why these adaptations may have evolved remain varied (and frequently controversial), but in many species they are considered to be a secondary sexual characteristic, with the bright, prominent colors of the bare skin aiding in courtship.

PAGES 76–77 Another possible evolutionary advantage of sparse or absent feathering in many bird species is thermoregulation. Unlike mammals, birds do not possess sweat glands. Especially for those living in hot-weather climates, bare legs and feet and sparsely feathered or naked areas on the head and face may help them keep from overheating. This blue-and-yellow macaw *(Ara ararauna)* is native to the tropical forests of Panama and South America.

 Parrots are not only among the most familiar of birds, they're also among the most widespread, found on every continent except Antarctica. Unsurprisingly, some of the world's most distinctive live on the islands of Oceania (Australia, New Guinea, and others), including the beautiful cockatoos (like this sulphur-crested, *Cacatua galerita*). The classification of cockatoos remains in flux, but it's thought that the ancestral species split away from the familiar "true" parrots (i.e., macaws, Amazons, and parakeets) about forty million years ago. Today, many of the more than twenty cockatoo species possess colors (such as white and black) rarely, if ever, seen in other parrot lineages.

 The physical changes wrought by artificial (or domestic) selection in the pigeons Charles Darwin bred included feathers that grew up instead of down (as in the Jacobin, left) and overlarge crops (the pouter pigeon, right). "I do not hesitate to affirm that some domestic races of the rock-pigeon differ fully as much from each other in external characters as do the most distinct natural genera," Darwin wrote in his 1868 book *The Variation of Animals and Plants under Domestication*.

 For thousands of years, humans have been breeding pigeons for food and other uses. Clockwise from top right: the scandaroon (or runt), a favorite of Darwin's and a breed once used for carrying messages; the Oriental frill, a royal bird once bred for the Ottoman sultans in Turkey; the nun, originally prized for its tumbling aerial displays; and the Hungarian giant house pigeon. Despite domestic pigeons' significant differences in size, plumage, bill length, and other features, Darwin believed that all were descended from a single common ancestor: the familiar rock pigeon. And he was right. In recent years, the genomes of more than forty pigeon breeds have been mapped, confirming his belief while also pinpointing the specific genetic mutations that cause many of the distinctive physical differences seen among various domestic pigeon breeds.

PAGES 84–85 Diversity among birds isn't found only in size, shape, habitat, plumage, and behavior. It begins far earlier. This collection of eggs (in the Great North Museum: Hancock, Newcastle upon Tyne, England), demonstrates some of that diversity as well. In nature, birds' eggs can range from those of the bee hummingbird (the size of a pea) to the 2.5-pound monstrosities laid by the ostrich. From the 1890s through 2009, the Great North Museum was called the Hancock Museum, after the great nineteenth-century British naturalists Albany and John Hancock, friends and supporters of Charles Darwin.

DIVERSITY IN COLD BLOOD

As animal groups whose history reaches back hundreds of millions of years, reptiles, amphibians, and fish embody the transience and vulnerability of species.

Some of this is clear from the fossil record. It's impossible to look at the skeleton of *T. rex* or the gigantic shark Megalodon and not be struck by the marvels that once inhabited the earth but do no longer.

And the vulnerability persists into the modern day. With some exceptions, it's very difficult to tell when a species has gone extinct. Yet it is easy to see how a combination of overfishing, pollution, and habitat destruction has decimated the populations of widespread ocean fish, from groupers to halibut to tuna. Extinction looms for them and others.

The threat to frogs and salamanders is even more dire, though less well known. Many amphibians live in fragile wetland environments that are under constant pressure from development, pollution, climate change, and other causes. In recent decades, they've also proven to be extremely vulnerable to a fungal disease that is considered at least partly responsible for the extinction of at least one hundred frog species.

Without renewed focus and attention, frogs—and other amphibians, reptiles, and many fish—could soon join the non-avian dinosaurs that linger only in Earth's memory.

PAGE 86 The mapping of octopus DNA has revealed that these fascinating creatures are extraordinary genetically as well as in other ways — such as having the ability to fit into small spaces of virtually any shape. With a genome that is almost as large and complex as a human's, the octopus possesses about half a billion neurons, more than fish, reptiles, amphibians, and even some mammals. Remarkably, most of these neurons are not found in its head, but down through its eight arms. As a result, each arm responds independently to threats and other stimuli, a skill it retains even when separated from the rest of the body.

OPPOSITE Islands are the birthplace of great diversity — and they're also where many of the world's most threatened species currently reside. (Island populations are rarely large to begin with, making them especially vulnerable to hunting or habitat destruction.) A case in point is Madagascar, home to lemurs, unique families of birds and plants, and a host of endemic reptiles, including the ploughshare tortoise or angonoka *(Astrochelys yniphora),* whose leg is pictured here. Listed as critically endangered by the International Union for Conservation of Nature, the ploughshare tortoise is estimated to number no more than four hundred individuals, and could be extinct in the wild in the next fifty years, along with many of Madagascar's other unique inhabitants.

PAGES 90–91 The five-toed foot and long, powerful tail of a saltwater crocodile *(Crocodylus porosus).* The first crocodylomorphs (the group that included the early crocodilians and their now-extinct relatives) included animals such as *Erpetosuchus,* which was small, terrestrial, and possibly bipedal. But after appearing in the Late Cretaceous Period, the true crocodiles have survived while hewing largely to the form we see today. Why have crocodilians endured so long without evolving into dramatically different forms? Like other big, aggressive predators (such as great white sharks), crocodilians rely on their size, strength, and sharp teeth to overwhelm their prey. This hunting technique is simple and effective, as its great longevity demonstrates.

PAGES 92–93 Crocodile eyes have evolved an assortment of specializations that make them suited to the animals' semi-aquatic environment, choice of prey, and hunting technique. The eyes are protected with a third eyelid, a membrane that slides across when the reptile submerges, while the eyeballs themselves can be drawn into the eye sockets during an attack. Meanwhile, a thin layer of guanine crystals (called the retinal tapetum) lies just behind the eyes. By reflecting light back through the retina, it intensifies the image, aiding crocodiles as they hunt in low-light situations. The reflection of the light cast by a flashlight on the tapetum creates the eyeshine seen in these and other animals.

PAGES 94–95 AND OPPOSITE Anoles (family
Dactyloidae) are among the most familiar of all lizards.
What is less widely known is that the anoles have been
found to evolve far more rapidly than was previously
thought possible in nature. For example, the species
Anolis sagrei (native to Cuba and the Bahamas) can
acquire longer hindlimbs in a single generation, giving
it the ability to climb more easily to evade predators or
pursue prey; similarly, when the species *A. carolinensis*
(native to the United States) was threatened by invading
A. sagrei, it developed larger toepads that facilitated
its ability to climb higher in foliage. These and other
unexpectedly rapid adaptations are forcing scientists
to rethink the conventional wisdom on the pace
of evolution.

ABOVE Wallace's flying frog (*Rhacophorus nigro-
palmatus*), native to Malaysia and Borneo, was first
described by its namesake, Alfred Russel Wallace, in
1869. Like a few other species of tree-dwelling frogs,
it has evolved membranes between its toes and flaps
of skin on its sides that, when spread, allow it to glide
long distances — as much as fifty feet — once it leaps
from a tree. Hence its common name: "parachute frog."

LEFT Natural selection has allowed some species to adapt and thrive in ecological niches that are not as accessible to their close relatives. For example, a unique oxygen-binding protein in their blood allows sperm whales to hold their breath underwater for up to ninety minutes. And most geckos, like this Tokay gecko (*Gekko gecko*), can stick to walls and ceilings as if glued there. In recent years advances in microscopy have explained how the lizards can accomplish this feat. It turns out that their foot pads are equipped with millions of setae, tiny hairlike structures with even tinier spatula-shaped tips. The vast number of contacts between the setae's tips and a vertical surface gives geckos' feet tremendous adhesive ability.

OPPOSITE Resembling a miniature dinosaur, and moving through its Australian desert haunts like a clockwork toy, the thorny devil (adorned with the wonderful scientific name *Moloch horridus*) is one of the strangest reptiles on Earth. It's also a living demonstration of how animals have evolved to survive in even the most forbidding environments. Water is the most precious commodity in the desert, and the thorny devil has a inimitable technique for taking advantage of the little that's there: It is able to wick moisture through tiny grooves in its body up into its mouth in a matter of seconds, in effect using any part of its body as a straw.

ABOVE AND OPPOSITE In many ways, despite their diversity (roughly 550 genera and 3,500 species known so far), snakes all resemble each other pretty closely. The real glory of snake diversity can be seen most clearly in their hunting techniques. For example, rattlesnakes and cobras quickly kill their prey with a venomous bite; while constrictors throw the coils of their body around their prey, gradually suffocating it with muscles that can exert twelve pounds per square inch of pressure, and then consume it whole.

PAGES 102–3 Until recently, the general scientific consensus suggested that all snakes descended from a four-limbed (and possibly semiaquatic) lizard ancestor, and that their skeletal structure reflects a kind of "reverse evolution" reflected by the simplification or loss of what are known as Hox genes, which govern the distinctions between the neck, trunk, lumbar, sacral, and tail regions in the backbones of limbed animals. But paleobiologists recently discovered that the expression of Hox genes in snake vertebrae is just as complex as it is in lizards. No loss or simplification has occurred, leaving the path of snake evolution an open question once again.

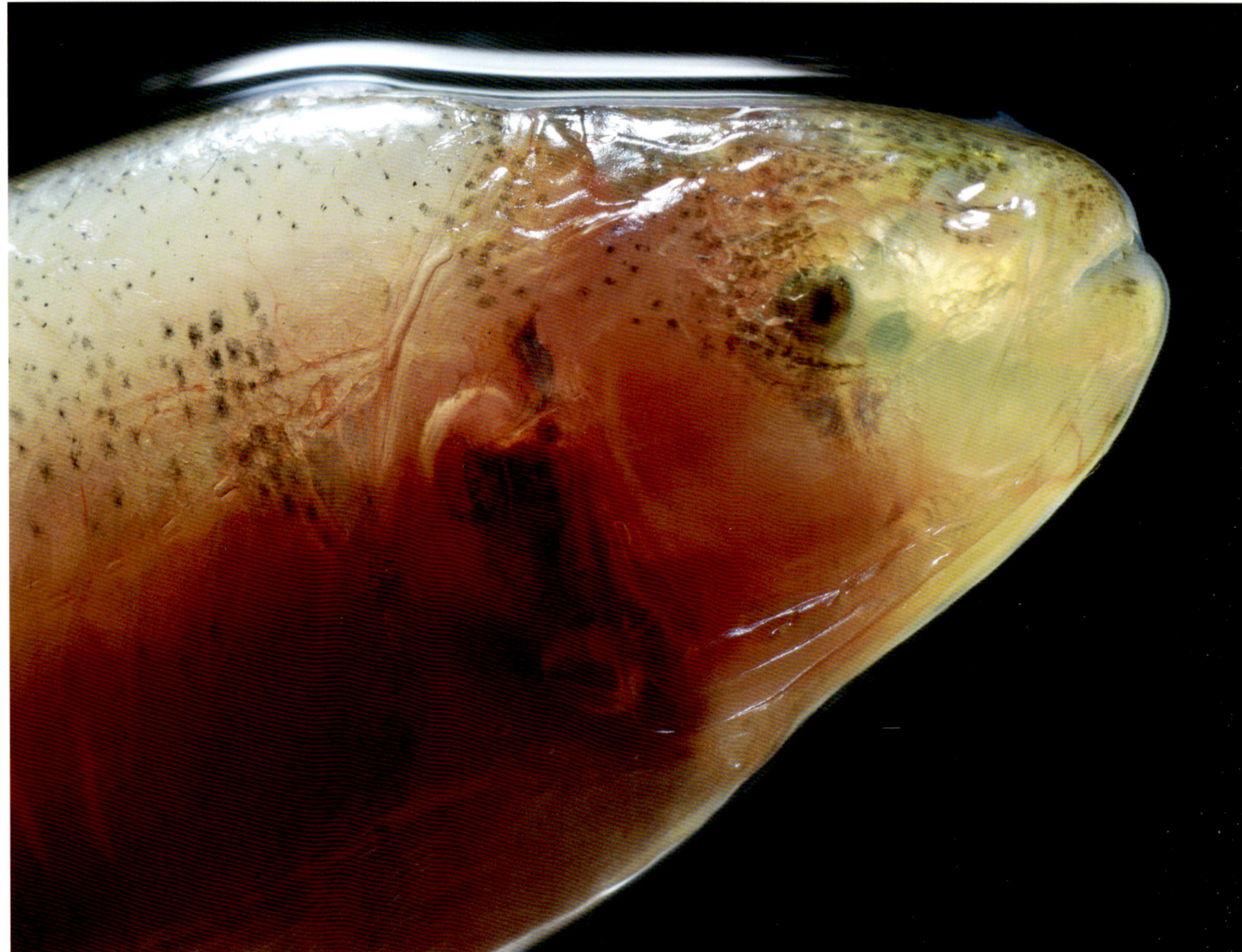

ABOVE How did the distinctive Mexican blind
cavefish (Astyanax mexicanus) lose not only its vision,
but its eyes, when adapting to life in pitch-black caves,
while its close relatives on the surface maintained
perfect vision? Researchers have found the presence
of "cryptic" or "standing" genetic variations in the
cavefish's sighted relatives (called the Mexican tetra,
despite being the same genus and species). These
variations caused no changes in sighted fish over
generations, but were "unmasked" after a population
of the tetras moved into dark caves, where eyes were
not needed. The unmasked gene allowed either bigger
or smaller eyes to appear in individual fish. Then
natural selection took over, selecting more energy-
efficient fish with ever-smaller eyes. The ultimate
result: the sightless cavefish found today.

OPPOSITE Such natural processes as changes in
climate, periods of drought, and extinction of food
sources can all affect the course of evolution by
natural selection. In recent centuries, a new major
factor has been human activity, whose impact appears
in some cases to be affecting other species' very
evolution. For example, a recent study shows that
intensive fishing with trawl nets may have created
a population of the largehead hairtail (Trichiurus
lepturus) that tends to be significantly smaller than
earlier populations at the same ages. (The smaller
fish can escape from the nets.) The ultimate effect
of these changes on the fish species' survival, if any,
remains uncertain.

 Unsurprisingly, given their fragility and lack of any skeletal structure, jellyfish are rarely preserved in the geologic record. That makes some recent finds of spectacularly well-preserved specimens in Utah's Marjum Formation (dating back to the Middle Cambrian Period around 505 million years ago) especially important. Helping clarify the origins of the Cnidaria — the phylum that also includes corals, and other aquatic invertebrate animals — these discoveries indicate that jellyfish may have evolved as long as seven hundred million years ago. They are the earliest known animals to possess true organs as well as possibly the first to attain the ability to swim actively rather than just float with the currents.

 Design experts have always looked to nature for inspiration. But in 2006 Mercedes-Benz became the first car company to create a vehicle based on the design of a fish. And not a sleek one, either: a squarish saltwater fish known as the yellow boxfish (*Ostracion cubicus*). The car designers believed that the boxfish's shape and an unusual, rigid bony carapace beneath its skin allowed it to reduce drag and aid stability. In 2015, however, researchers revisiting the issue discovered that the boxfish's shape and odd skeletal structure do not accomplish either goal. Instead, its speed and agility depend on something not reproducible in an automobile: well-coordinated movements of its fins.

ABOVE A microscopic view of sharkskin, a fascinating evolutionary development. Unlike most other fish, sharks do not possess air bladders to help them stay afloat. Instead, to keep from sinking, they must swim at all times, which is an energy-intensive chore. They have evolved features on their skin that reduce drag and make moving through the water more efficient. These tiny teethlike structures, called denticles, may also aid in propulsion by creating a small whirlpool that spins just in front of the shark as it swims.

OPPOSITE Every species evolves its own mechanisms for reproduction. Many mammals, for example, bear only one or two young at a time, but protect them fiercely. Other animals, including the salmon that spawned these alevin (newly hatched fish still dependent on their yolk sacs for survival), rely on different techniques. A single female salmon can lay thirty-five thousand eggs before she dies soon after. Once the alevin hatch out, the species depends on the sheer force of numbers for survival.

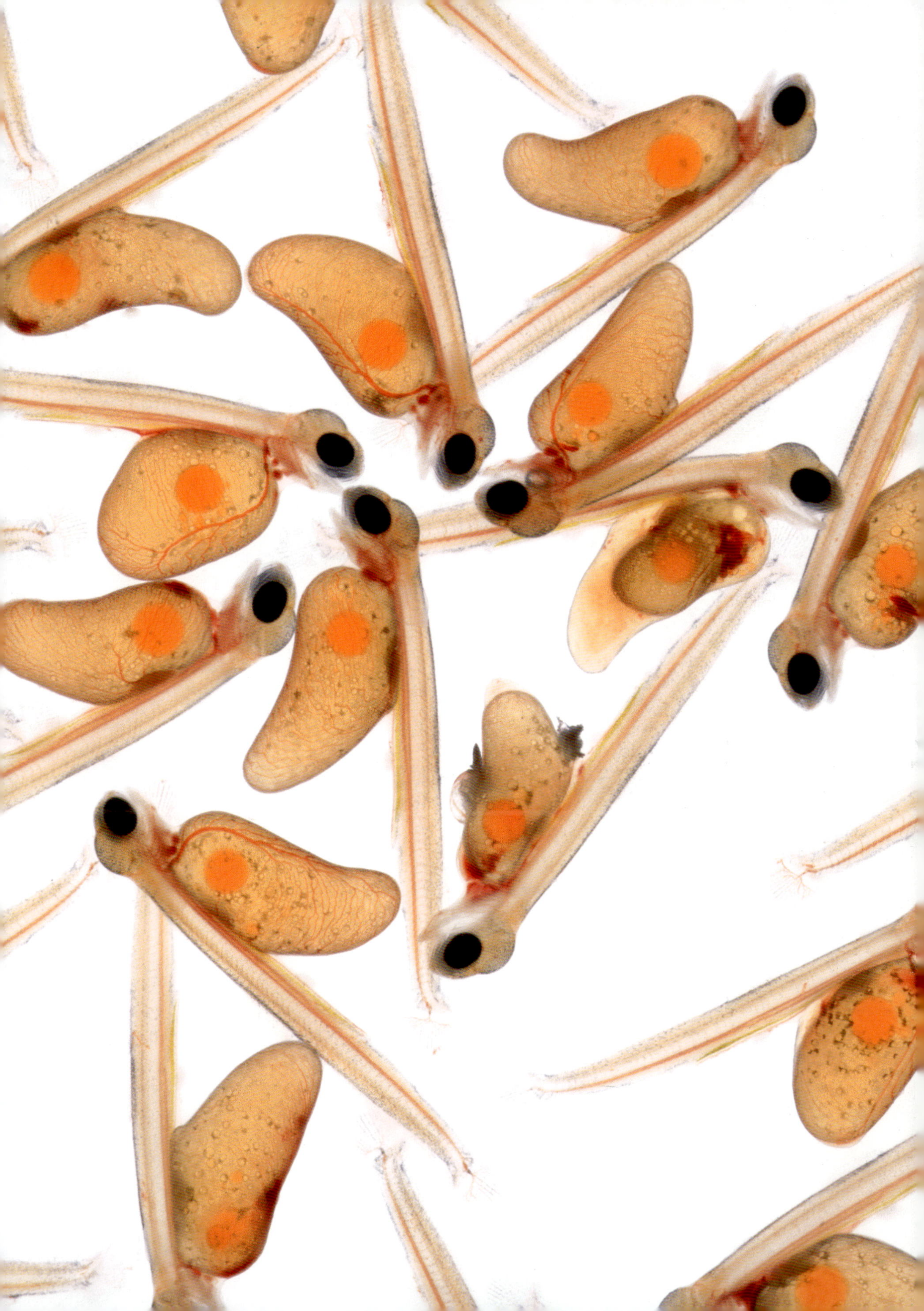

PLANTS
EVOLUTION BLOOMS

While plants have rarely numbered among the best-known examples of evolution, their history (and especially that of angiosperms, or flowering plants) has been replete with examples of survival in the face of harsh challenges. Since they first moved from aquatic environments onto land about 450 million years ago, plants' ability to spread—to evolve unparalleled means of propagating themselves effectively—has been one of the great stories of adaptation.

Consider the act of pollination. Unlike almost every other form of life, individual plants have practically no ability to move in order to find a mate—yet in most plants, pollination is as crucial to producing offspring as direct, one-on-one fertilization is to animals. Plants have responded to this challenge through the evolution of a breathtaking array of techniques, including temptation, trickery, and transportation.

The same mix of approaches holds true with the output of successful pollination: the seed, which often can best ensure the survival of a species if dispersed far from the parent plant. Meanwhile, the additional strategies that plants have evolved to survive in a world filled with hungry herbivores—ranging from mimicry to the manufacture of complex toxins—are just as eye-opening.

PAGE 112 Orchids (family Orchidaceae) are a spectacularly large and widespread group of flowering plants, with at least twenty-five thousand species described so far. Given how many orchids are found in forest canopies, this *Paphiopedilum superbiens* (endemic to the highlands of central Sumatra) has an unusual habitat. It grows only in leaf litter.

BELOW AND PAGES 116–17 The earliest plants developed in water and emerged on dry land about 450 million years ago, roughly the same time that the first animals did the same. At first, many land plants evolved spores to propagate themselves, as ferns still do today. About 360 million years ago, through a succession of mutations, plants called gymnosperms began to create two types of spores: male and female, sperm and egg. If the two came into contact after being released by the plant, fertilization could take place. This was the beginning of an evolutionary leap forward that coincided with an explosion in plant diversity across the planet, including the appearance of angiosperms (flowering plants) and their wide diversity of seeds and fruit.

OPPOSITE Look at a fern today, and you're seeing an ancient survivor, both of the great extinctions that have periodically struck the planet and of competition from flowering plants, which evolved about 120 million years ago. Studying how some ferns managed to survive, researchers came up with a possibility: Those ferns gained an advantage by possessing a protein (called a neochrome) that allows them to absorb and use light in the red spectrum, not just the blue spectrum absorbed by flowering plants. Strangely — and no one knows how yet — the gene for the neochrome molecule does not appear to have evolved in the fern itself, but somehow to have been transferred in from a plant called a hornwort. The dark spots on this fern are composed of its spores.

OPPOSITE One very early, successful step in plant evolution was the development of pollen (here, visible on a dying blossom). Fossil specimens of seed ferns with pollen attached date back to the Late Carboniferous Period (about three hundred million years ago). The evolution of pollen conferred a dramatic survival advantage: While the spores of more primitive plants required water to spread, pollen could be carried by the wind, a step that opened up vast new areas for colonization.

PAGES 120–21 Flowering plants and trees have evolved a nearly incalculable variety of techniques to aid survival and propagation. The seeds of some orchids, for example, weigh only 1/35,000,000th of an ounce, sailing through the air like dust particles before lodging in the canopies of tropical trees. On the other hand, the seed of the coco de mer palm can weigh up to forty pounds and can float unharmed across vast stretches of ocean waters to colonize new coasts.

PAGE 122 Meet the carnivorous Venus flytrap (*Dionaea muscipula*). Native only to wetlands in North and South Carolina, it's famous as one of the few carnivorous plants that uses sudden movements to trap its prey. As if this weren't impressive enough, researchers have recently made another remarkable discovery: Venus flytraps can count. An open trap will not close unless its trigger hairs are touched twice within twenty seconds. And even once it closes, the trap will not secrete the enzymes it uses to digest its prey until it is triggered a third time. Beyond that, the number of times the hairs are triggered corresponds with the amount of digestive enzymes released. From start to finish, this is a memorable story of evolutionary adaptation.

PAGE 123 A sundew (family Drosera), one of nearly two hundred species of carnivorous plants that occur on every continent except Antarctica. Like Venus flytraps, sundews need more nutrients than the poor soil of their wetland habitats can provide, so the plants have evolved an innovative way to augment their diet. Their leaves come equipped with tentacle-like glands that produce a sweet, sticky fluid called mucilage. When an insect, drawn to what seems like a feast, touches one of the tentacles, the leaf curls to ensnare it. The insect, trapped in mucilage, soon dies of exhaustion or asphyxiation, and is digested by enzymes secreted by the sundew.

PAGES 124–25 *Prosthechea cochleata*, an orchid found in the wild in Central and South America and even in south Florida, is commonly cultivated for its uniquely shaped, long-lasting flowers. Orchids boast the most diverse group of pollinators of any plant family, including a host of birds, bees, wasps, butterflies, and moths, and even a species of cricket. The techniques some orchids use to attract pollinators are glories of evolution. For example, the genus *Ophrys*, the "bee" or "fly" orchids, mimics female bees so markedly (while also emitting intensely attractive scents) that male bees pollinate them in the midst of frustrated mating attempts. And one species, the South American sack-shaped catasetum *(Catasetum saccatum),* which intrigued Darwin himself, actually hurls its sticky pollen sacs at insects when they touch a part of the flower.

ABOVE One of the orchid's many survival techniques is seen in this Disa orchid *(Disa uniflora),* which lives on the frosty, windy heights of South Africa's Table Mountain and nearby peaks. Rather than entrusting its pollen to the breeze, the Disa has evolved fused male and female parts — the white column in the center of the flower. As a result, a single visit from the butterfly that pollinates it can allow the plant to produce fertile seeds. Disa orchids come in many varieties and are popular with collectors.

OPPOSITE Many orchid species have extremely small ranges. For example, this *Paphiopedilum rothschildianum* (Rothschild's slipper orchid) grows only in the rainforests on the slopes of Mount Kinabalu in northern Borneo, between sixteen hundred and four thousand feet above sea level.

 "I was never more interested in any subject in all my life than this of Orchids," Charles Darwin wrote to his friend the botanist Joseph Hooker. Darwin's most famous connection to these diverse and widespread plants involved *Angraecum sesquipedale* (now commonly known as Darwin's orchid), a species from Madagascar. Studying it, Darwin understood that the species could be pollinated only by a moth with a proboscis far longer than any yet seen on Earth. A variety of "experts" ridiculed Darwin's prediction, published in 1862, that such a moth *must* exist somewhere on Madagascar. (Ridicule was an experience he must have been used to by then.) It wasn't until 1903, more than two decades after Darwin's death, that collectors found the exact moth that he'd said was out there: *Xanthopan morganii*, originally given the subspecies name *praedicta* in recognition of the great man who predicted its existence.

 The Devonian Period (416–358 million years ago) is known as the Age of Fishes, and it did witness an extraordinary surge in fish diversity — including early sharks and lobe-fin fishes, the ancestors of the first limbed animals to walk on land. But the evolution of plants and trees at the time was equally dramatic, as the first forests began to spread across the land, and ferns and grasses, flowers, and other seed plants also appeared. Seeds and their evolution do not get the attention that more colorful examples of the phenomenon receive — whether it's the beaks of finches or the pollination of orchids. In many ways, though, life on Earth's surface as we know it rests on plants' ability to colonize nearly every inch of what was once a barren, rocky planet. And that ability would never have developed if seeds, sturdy structures capable of traveling great distances unharmed — even inside the stomachs and intestines of birds and other animals — had not evolved.

INSECTS
ADAPTATION IN ABUNDANCE

Without the omnipresence of insects—and other arthropods, such as spiders—in all of our lives, we might have had to wait much longer before someone explained evolution. Both Charles Darwin and Alfred Russel Wallace grew up enamored of nature, and the nature they encountered most frequently was bugs. Darwin especially was notorious for his love of beetles as a child, at least once apparently finding his hands so full of specimens that he attempted to carry one home in his mouth!

If arthropods, especially the insects, represent one thing most vividly in the story of evolution, though, it is not only abundance: It is mystery. While scientists have a rough idea of the number of existing species in some other groups (mammals, reptiles, etc.), no one—not even the most knowledgeable entomologist—truly has a clue as to how many insects are out there.

Some experts estimate that there is a mind-boggling total of one million species and others guess that the real number is likely to approach thirty times that many. No doubt, insects are by far the most widespread and abundant example of evolution on Earth. And—of course, in the species as yet undiscovered and the stories they have to tell us—the most mysterious.

LEFT Like the Madagascan sunset moth (pages 142–43), this jewel scarab beetle (genus *Chrysina*) uses structural iridescence instead of pigment to create its magnificent hues. The microscopic structure of its wing case causes light to be reflected only in certain spectra, allowing a species to gleam green, blue, or other bright colors. Like the duller-colored sacred scarab beetle of Egypt, those in the genus *Chrysina* (native to the Americas) currently belong in the gigantic family Scarabaeidae, which includes over thirty thousand named species.

OPPOSITE A tiny, fragile butterfly egg. To depend for survival on such a vulnerable vehicle, an easy morsel for a predator, would seem like an unlikely path to evolutionary success. But butterflies and moths first appeared during the Jurassic Period, perhaps 190 million years ago, and continue to thrive today. No one has any idea how many species exist right now, but estimates range from about 180,000 (17,500 butterflies, the rest moths) up to a half-million or more. And every new individual of every species begins with an egg much like this one.

PAGE 132 Asked what could be deduced about the Creator by the study of life on Earth, the prominent British scientist J. B. S. Haldane (1892–1964) supposedly replied, "An inordinate fondness for beetles." Although the quote is likely apocryphal, there is no doubt that evolution at least has favored the insects, and among them most of all the order Coleoptera, which now encompasses about four hundred thousand species. Scientists' estimates of the number of beetles that remain to be discovered vary widely, but even the lower figures are a testament to the coleopterids' evolutionary success.

PAGES 138–39 While described butterfly species number only around 17,500, the moths — their fellow members of the order Lepidoptera — are far more diverse. Among the roughly 160,000 species described so far, most are small and easy to miss. But some are the opposite, like this stunning Atlas moth *(Attacus atlas)*, from Southeast Asia and across the Malay Archipelago, where Alfred Russel Wallace found and described it. Along with its large size, the Atlas has a special technique for giving potential predators pause: It brandishes its yellowish upper wingtips, which, with their blue-black pattern, resemble the head of an angry cobra. The similarity is so striking that in parts of China, the Atlas is known as the "snake's head moth."

PAGE 140 A female Rajah Brooke's birdwing butterfly, *Trogonoptera brookiana*. Wallace, who described the species in 1855, named it for James Brooke, the eccentric British adventurer who was the colonial head of the state of Sarawak in Borneo. While waiting out the rainy season in Sarawak that same year, Wallace wrote his paper "On the Law which has Regulated the Introduction of New Species," which laid out the beginnings of his theory of evolution.

PAGE 141 Among the most noticeable butterflies in the tropical forests of the Americas are the "eighty-eight" or "numberwing" butterflies, gaudy species belonging to one of three genera: *Diaethria, Perisama,* and *Callicore.* (These photos show the upper- and underwings of one typical species.) Scientists have theorized that their showy underwing patterns, filled with dots and lines, may both startle and confuse potential attackers. The common name of these butterflies has a simpler explanation: In some species, the black-and-white underwing patterns resemble the numbers eight or zero.

OPPOSITE AND ABOVE The process of butterfly metamorphosis, in which a wormlike caterpillar hatched from an egg changes into a chrysalis and then into a winged adult, seems almost miraculous. This series of transformations was once considered unlikely enough that (while in Chile on the *Beagle*'s voyage) Charles Darwin encountered a German naturalist who'd been arrested for heresy — merely for stating that the caterpillars in his care might become butterflies! The heretical chrysalises here come from a birdwing butterfly native to Cambodia. Their range of colors partly reflects the kind of leaves that the caterpillars ate before beginning their metamorphosis.

First described in 1773, this beautiful lepidopteran (above left: top, above right: bottom, opposite: detail) endemic to Madagascar was first thought to be a kind of swallowtail butterfly. Further research, however, showed that it's actually a moth in the widespread family Uraniidae. Since it is large, noticeable, diurnal, and migratory, the Madagascan sunset moth would seem to be a prime target for birds and other predators — but it has an effective defense: It is toxic, and its bright colors, rather than being an invitation, serve as a warning. The moth's iridescence (and that of many other insects, as well as some birds) comes not from pigment, but through microscopic structures in its wings. These structures refract light in such a way that the wings' surface flashes bright orange and green or dull gray-brown, depending on the angle at which the light strikes it.

PAGES 144–145 The result of evolutionary adaptation can be startlingly elegant. Case in point: protective coloration or camouflage in the Indian or Malayan leafwing butterfly *(Kallima paralekta)*. (Despite its common name, the species is actually endemic to Sumatra and Java in Indonesia, where Alfred Russel Wallace encountered and wrote extensively about it.) The upper surface of the male's wing features a relatively straightforward pattern of blue and orange — but when its wings are closed, it suddenly resembles an ordinary brown leaf.

ABOVE The genetic mutations leading to gynandromorphy — the possession of male and female sexual characteristics by a single individual — occur throughout nature, appearing in crustaceans, birds, and other animals. It is especially noticeable in butterflies (such as this birdwing, *Trogonoptera trojana),* because many butterfly species already show sexual dimorphism — visible differences between male and female. This birdwing shows one bright "male" wing and one duller "female." The appearance of gynandromorphs among butterflies has long been recorded. In his autobiography, the writer and lepidopterist Vladimir Nabokov recalled one he'd caught as a child in Russia, and more recently, a gynandromorphic great Mormon butterfly made headlines after emerging from a chrysalis during a 2011 butterfly show at the Natural History Museum in London.

OPPOSITE For obvious reasons — above all, their lack of any kind of skeleton and overall fragility, especially in the wings — butterflies and moths don't appear often in the fossil record. As a result, only recently have experts begun to agree that the first butterflies appeared during the Jurassic Period, perhaps 190 million years ago. (This allows us to imagine a butterfly perching on the nose of a *Dilophosaurus* or other dinosaur out of *Jurassic Park,* as many do on the noses of river turtles today.) This swallowtail moth *(Lyssa zampa)* is found in China, Singapore, Borneo, and elsewhere.

PAGES 148–49 The 550 or so species of swallowtails (family Papilionidae, such as this *Papilio lormieri),* include some of the most familiar of all butterflies. Most are large and gaudy. Even their caterpillars are distinctive: huge, fleshy, brightly colored, and in some cases possessed of huge eyespots that scare away predators. As if that's not distinctive enough, every papilionid caterpillar comes equipped with a special gland that can suddenly emerge like a pair of antlers from behind its head and emit a toxic liquid.

PAGES 150–51 Like scarab beetles (page 134),
these two goliath beetles are members of the family
Scarabaeidae, joining four other species in the genus
Goliathus. The six species are among the largest
beetles in the world, with one reaching more than
four inches in length. They are all native to Africa's
tropical forest, where they subsist on a diet mainly
made up of tree sap and fruit.

RIGHT AND OPPOSITE An array of beetles. It seems
that the earliest beetle-like insect evolved during
the Carboniferous Period, perhaps three hundred
million years ago, though true beetles did not start
appearing until the Triassic, about 220 million years
ago. Between then and now, they've evolved to fill
niches in nearly every environment on Earth except
for saltwater and the extreme polar regions.

Among the most spectacular of all insects are the rhinoceros beetles, which belong to the family Scarabaeidae, the widespread and abundant scarab beetles. By beetle standards, they can grow to enormous sizes, some reaching six inches in length. Even with their fearsome "horns," they are slow-moving and unaggressive, avoiding predators by relying on their nocturnal habits, concealing them-selves in vegetation, and (in some species) making loud squeaking sounds by rubbing their abdomens against their thick wing cases.

OPPOSITE The mango longhorn beetle (*Batocera rubus*) is a member of a typically extravagant family of beetles (Cerambycidae) currently comprising at least twenty-five thousand species. The feeding habits of many species of longhorn beetle — especially their larvae — demonstrate how evolutionary adaptations aided by the relatively recent ease of travel provided by human conveyances have allowed the beetles to become serious pests on fruit and other crops. The mango longhorn beetle preys not only on mangoes, but on figs, apples, and other fruit, while the larvae of its relative the Asian longhorn beetle (*Anoplophora glabripennis*) threatens entire forests in North America and elsewhere.

ABOVE To reach their stunning level of diversity, beetles have managed to fill a wide variety of ecological niches, from arid deserts to flooded rainforests and even beneath the surface of ponds and lakes. Their diets are equally varied: While many beetles eat only plant materials, others (like the Dermestid beetles used by taxidermists and museums to clean specimens) feast on flesh, and still others (most famously the dung beetles) exploit their namesake.

PAGES 158–59 The dead leaf mantis (genus *Deroplatys*, left) and true leaf insect or "walking leaf" (family Phylliidae, right) provide two examples of how color and shape — both of which aid survival through camouflage — have been naturally selected for countless times throughout evolutionary history. The skillful mimicry of leaf insects (native to South and Southeast Asia and Australia) enables them to hide almost invisibly on branches or in leaf litter.

ABOVE AND OPPOSITE Stick insects, members of the order Phasmatodea, are closely related to mantids and cockroaches. The more than twenty-five hundred species occupy a wide variety of habitats, but reach their greatest diversity in tropical and subtropical forests. Stick insects number among the largest insects on Earth, with some species exceeding twelve inches in length. Unlike their mantid relatives, they eat an entirely vegetarian diet, most often leaves.

It's easy to mistake a crane fly for a nightmarishly large mosquito. Aside from both being insects of the order Diptera (true flies), however, the two groups differ markedly. Perhaps most importantly, crane flies (family Tipulidae) don't feed on the blood of humans or any other species. In fact, though their aquatic larvae eat other small invertebrates, many adult species don't live long enough to eat at all. Typically for many groups of insects, the classification of crane flies remains complex and tangled, as do estimates of the ultimate number of species. Currently more than fifteen thousand have been identified.

ABOVE Widespread, adaptable, the ultimate survivors: Cockroaches may be the most familiar — and least loved — of all insects. Their relatives have been found in fossil beds from Carboniferous sites (about 320 million years old), while the first true cockroaches appear in the fossil record in the Triassic Period. Over time the most familiar species, like the American cockroach *(Periplaneta americana),* have evolved the ability to run at astonishing speeds — up to two miles per hour in short bursts — and compress their bodies to fit into cracks that measure just a third of their standing body height. (But everyone who's ever seen a cockroach racing across the kitchen floor already knows all this.)

OPPOSITE Like insects, spiders (part of the same phylum, Arthropoda, but not closely related) have a long and complicated evolutionary history stretching back more than three hundred million years. Especially fascinating have been scientists' efforts to untangle the evolution of venom, particularly its genetic basis, which varies so much from species to species. Even individual groups, such as the Australian funnel web spiders or the black widow and its relatives, have evolved such complex cocktails of venoms that scientists are still working to identify their individual chemical constituents.

MAMMALS
A WEB OF CONNECTION

Scientists agree that natural selection is the driving force behind the evolution of life on the planet. Inevitably, however, many uncertainties remain about the details of the process, and—especially in an era of intensive study with powerful tools ranging from scanning technology to DNA sequencing—it often seems as if new questions are being asked at least as quickly as old ones are answered.

One recently revived debate has to do with the pace of evolutionary change. Charles Darwin believed, and fossil records have seemingly indicated, that vast amounts of time are required for new species to evolve from existing ones. On the other hand, Darwin himself (and anyone else familiar with the breeding of dogs, pigeons, or flowers) knew that, with human involvement, artificial selection can cause significant physical changes—and the creation of new breeds—in just a few generations.

Now, scientists are discovering that rapid evolution seems possible, if not frequent, in nature as well. Two examples: The appearance in the past few decades of the eastern coyote (whose immediate ancestors include wolves and dogs) and the recent discovery of a grizzly and polar bear hybrid that is capable of producing fertile offspring.

What constitutes a species? We do not yet know for sure. What we do know is that the study of evolution is itself constantly evolving.

PAGE 166 Despite their extreme differences, all dogs are the same species, *Canis lupus familiaris* or *C. familiaris*. While puzzling out the origins of the domestic dog (such as the English setter, here), scientists are also grappling with a far larger question that dogs and their closest relatives have brought into the open: What constitutes a species? As anyone in the eastern United States knows, the region has seen an invasion of coyotes in recent decades. Only they're not really coyotes. DNA evidence has shown that these animals are, in fact, a mix of western coyote (*C. latrans*), wolf (*C. lupus*), and dog. They have interbred over just a handful of generations to produce fertile offspring, something previously believed possible only within a single species. In all likelihood, this interbreeding has produced either a newly emerged species (the eastern coyote, or coywolf), or merely another facet of what is, in fact, a single species that includes all three of those that contributed to its DNA.

OPPOSITE Just as it has with dogs and pigeons, artificial selection managed by human breeders has led to a dramatic variety of horses (*Equus caballus*), ranging from huge draft horses to fine-boned Arabians to this Shetland pony, native to Scotland's Shetland Isles. Although short, stocky Shetland ponies have been employed for carrying peat, plowing fields, and other heavy-duty uses for countless generations, exactly when and how they came to the islands remains unclear. Perhaps they were brought by Norse explorers or by sailors in the Spanish Armada. Or perhaps the very first settlers of the Shetlands brought these trustworthy, hardy, and even-tempered ponies with them thousands of years ago.

PAGE 170 "One of my chief objects in coming to stay in Simunjon [on Sarawak]," Alfred Russel Wallace wrote in his book *The Malay Archipelago*, "was to see the Orangutan (or great man-like ape of Borneo)." During his visit, he observed the way the apes (called "mias" by the local people) moved through the trees, commenting, "It is a singular and most interesting sight to watch a Mias making his way leisurely through the forest. He walks deliberately along the branches, in the semi-erect attitude which the great length of his arms and the shortness of his legs give him." Wallace did more than watch (and collect) this magnificent primate and other local fauna while in Sarawak: During a long, solitary rainy season there he wrote "On the Law which has Regulated the Introduction of New Species," his first statement on evolution and natural selection.

PAGE 171 The hands of a gorilla. Despite the fact that humans are bipedal and gorillas usually walk on all four limbs, the two species' hands are closer in proportion to each other than to those of any other ape, including *Homo sapiens'* closer relative, the tool-using chimpanzee. According to a recent study, one reason for this may be tied to the fact that both gorillas and humans use their hands (cupped in gorillas, clenched into fists in humans) for ritualized aggressive display and attack, unlike other apes. The similar structure therefore may be an example of parallel evolution.

 In many cases, the varied pathways of evolution are subtle and hard to see. In others, not so much. Behold the echidna (family Tachyglossidae), four species of which are found in Australia and New Guinea. Along with the better-known platypus, they are the world's only surviving monotremes, mammals that lay eggs. Fossils allowing scientists to detail the steps in monotreme evolution remain elusive, but the oldest fossil echidna dates to about seventeen million years ago, during the Miocene Epoch. Like the modern-day species, it had a toothless skull and a skeleton designed for digging, suggesting that it filled niches similar to those of today's ant- and grub-eating species.

PAGE 174 A fetus of a donkey exhibits many of the features that distinguish the family Equidae, which also includes horses and zebras. Astonishingly, scientists recently found a powerful tool to help untangle the complex evolutionary history of the family: a seven-hundred-thousand-year-old colt leg bone that had been frozen in the tundra of Canada's Yukon Territory. The bone's excellent condition allowed scientists to map the ancient horse's entire genome. (It remains by far the oldest DNA yet sequenced.) Among the researchers' many findings from this scientific trove: Horses, donkeys, and zebras appear to have evolved from a common ancestor 4 to 4.5 million years ago, or twice as long ago as previously thought.

PAGE 175 The naked mole rat (*Heterocephalus glaber*), along with the Damaraland mole rat, is one of the only two eusocial mammals on Earth. (*Eusocial* means they live in a hive-like community or colony, as ants and bees do.) Researchers have discovered something else that makes mole rats unique among mammals: They possess a genetic mutation that allows them to produce a sugar that prevents their cells from forming tumors. At least partly as a result, mole rats can live more than thirty years, compared to just a few for most other rodents, and show few signs of aging during their long lives. Long before the discovery in the 1980s that naked mole rats were eusocial, Charles Darwin wrestled with how natural selection would work in a population where only a single queen bred. He concluded, rightly, that the evolutionary process would largely occur between families or hives, not within a single colony.

ABOVE AND OPPOSITE The evolution of bovids
(family Bovidae, a large, widespread group of cloven-
hoofed, ruminant mammals that includes everything
from African buffalo to tiny antelopes to sheep, goats,
and domestic cattle) is a complicated subject. Even
the history of horns and antlers (such as those of the
urial, above, and blackbuck, opposite) remains unclear,
with expert opinion divided on the reason that horns
occur in the males only in some species and on both
sexes in others. One theory: Since females of territorial
species often have horns, while nonterritorial females
do not, horns may have evolved for defensive purposes
in the former. Conversely, in nonterritorial females
(which, not being bound to a specific site, can rely more
on camouflage to evade predators), similar mutations
have provided no selective advantage, and therefore
have not persisted.

PAGES 178–79 A beast whose signs were seen often
by Alfred Russel Wallace, the hairy-flanked, two-horned
Sumatran rhinoceros *(Dicerorhinus sumatrensis)* is
now one of the most endangered mammals on Earth.
It also demonstrates how geographic isolation
leads to the development of new species. During a
time when sea levels were low, ancestral rhinos spread
eastward from their mainland African and Asian
homelands. Some individuals made their way to what
became the islands of Sumatra and Java. Becoming
isolated when water levels rose, they eventually
evolved into the distinctive species that exists today.

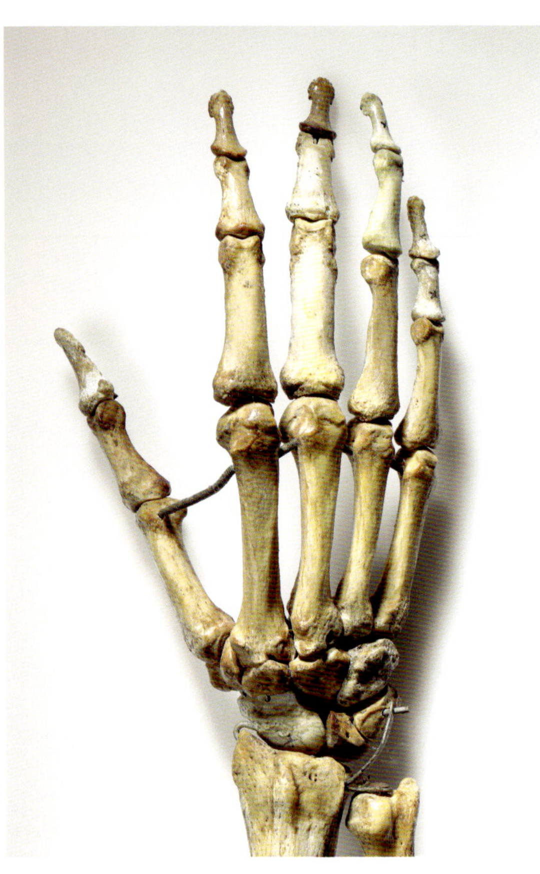

 The relentless ongoing process of natural selection has led from the earliest ancestral mammals (one candidate being a shrewlike, egg-laying mammal living in the Late Triassic Period) to the marvelous diversity we see today. With its opposable thumb, relatively short fingers, and wide, flat palm, the human hand (set free from having to aid in locomotion) is well suited for grasping, undertaking an enormous variety of precision tasks, and making a fist in order to throw, threaten, or deliver a punch.

 The last common ancestor of whales and humans lived soon after the great extinction event that brought an end to the Late Cretaceous Period about sixty-six million years ago. Yet some skeletal features still show the similarities of common ancestry, such as the "toes" possessed by both groups, as shown by this fossil whale hindlimb. In a modern whale, the tiny hindlimbs are hidden within the flesh, while the "hands" evolved into powerful flippers to make it peerlessly suited to its marine environment.

 Everyone who's looked knows that domestic cats' eyes have vertical slit pupils while those of humans and most other animals are round. A recent study identified an association between slit pupils and predator species that ambush their prey and hunt extensively at night, as small cats do. These pupils, the researchers suggested, give a clear view of the distance to a specific object (such as a prey animal) while letting the greater panoramic view go out of focus. This could provide a selective advantage by helping the cat accurately gauge its pounce in low-light conditions. Predators that chase down their prey and tend to be active both day and night (such as lions and other large cats) are more likely to have round pupils.

PAGES 184–85 While artificial selection has allowed breeders to create domestic dogs in an eye-popping variety of sizes and shapes, scientists continue to puzzle out where they originated — and when. It is commonly accepted that dogs are descended from gray wolves, but experts' recent estimates on when the species diverged range from eleven thousand to as long as forty thousand years ago. In addition, new data (including the sequencing of the genome of a thirty-five-thousand-year-old wolf) reveal the presence in dogs of genetic material not only from European gray wolves but also from the extinct Taymyr wolf of Asia.

ABOVE AND OPPOSITE In the process of evolution through natural selection, mutations are favored when they aid in an individual's — and, ultimately, its species' — survival. No such rules must be followed during artificial, or domestic, selection, in which breeders (whether of orchids, fish, pigeons, or dogs) can select for features based on a personal goal, not the individual's ability to survive in nature. The English bulldog's build and skull demonstrate this, while also making clear why the breed suffers from a variety of respiratory, cardiac, skeletal, and other health problems.

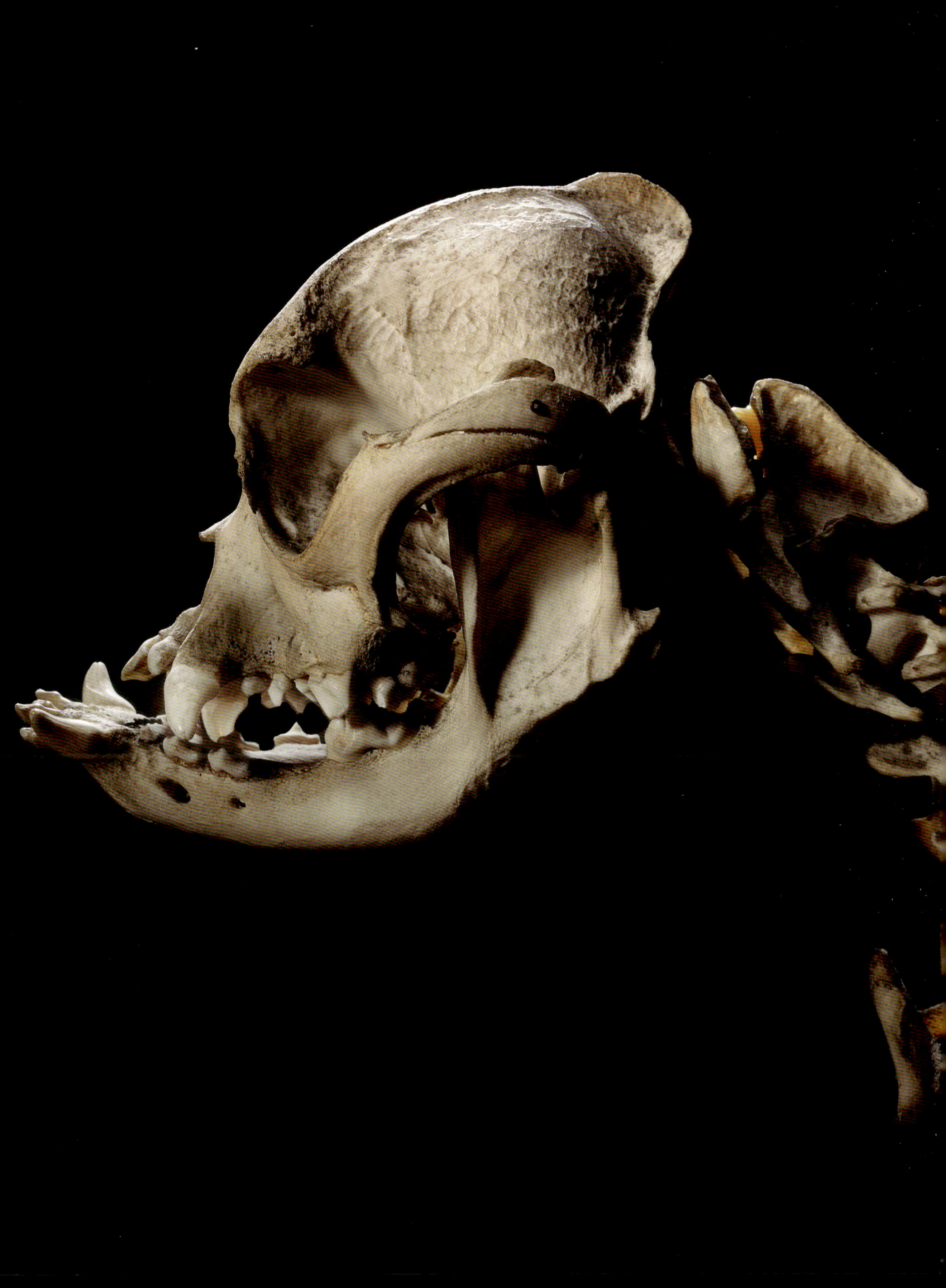

PAGES 188–91 According to the scale of typical evolution, most dog breeds have been in existence for only an eyeblink of time. But by the standards of human history, many are practically ancient, demonstrating both the longevity and the tightness of the bond between humans and canines. For example, the bloodhound (page 190) came from hounds bred in Europe more than a thousand years ago; giant boarhounds closely resembling modern Great Danes (above) were bred in ancient Greece; and the Afghan hound (opposite), though having a less well-defined past, has a centuries-long history in Afghanistan and surrounding areas. The bichon frise (page 191), unlike the others, was bred for companionability and a comfortable life in royal courts. It dates back at least to the thirteenth century.

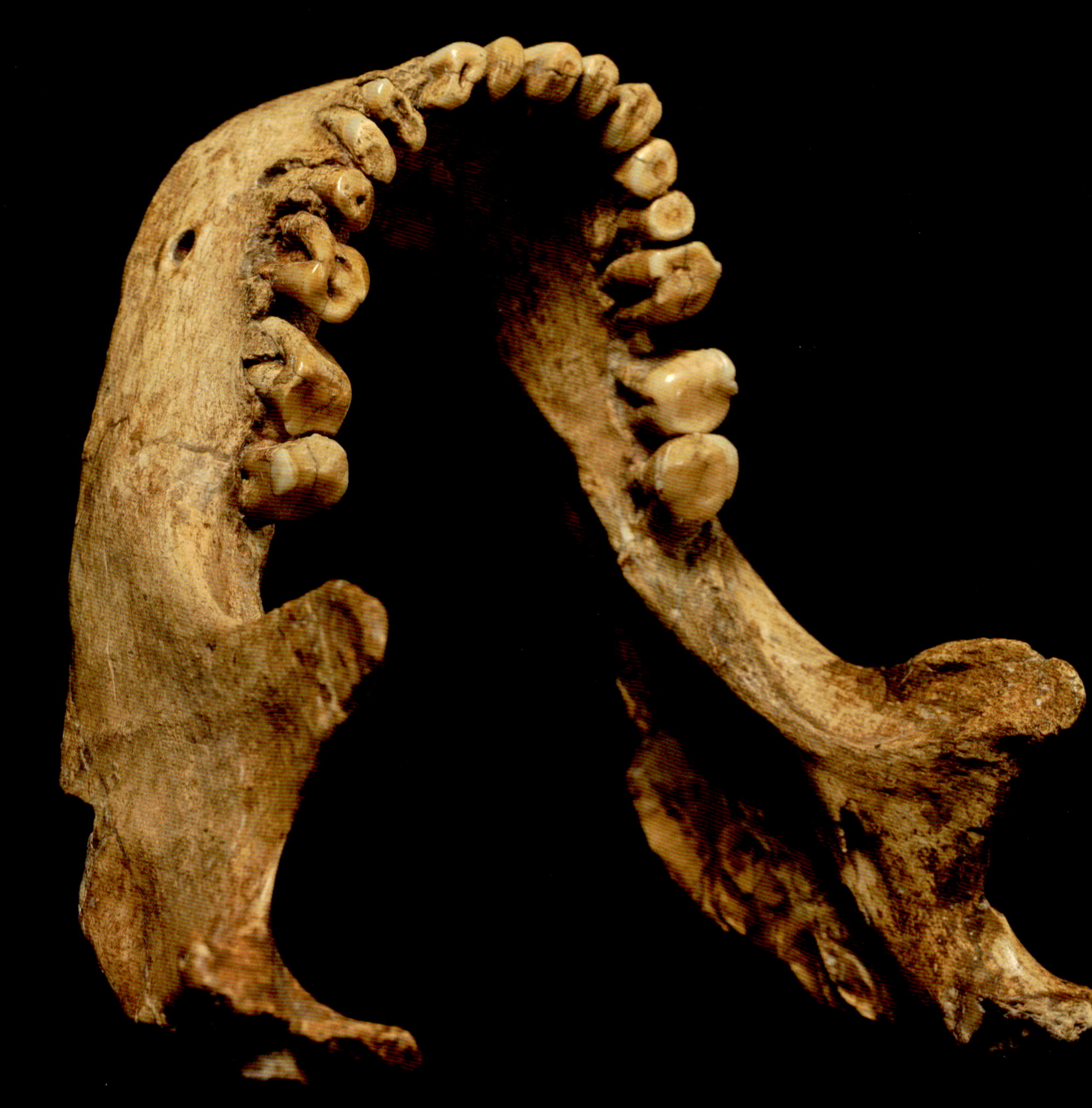

HUMAN ORIGINS

Just a century ago, the concepts that humans and modern apes share common ancestors and that the earliest humans arose in Africa were endlessly controversial. Today, however, all but a few acknowledge this established history. Still, it feels like every year brings fresh fossil evidence or a new laboratory analysis that upends previously accepted wisdom on the precise details of the origins of our species.

Intense searching and new research tools are driving the study of human origins in unpredictable directions. For example, the ability to sequence species' entire genomes has allowed scientists to determine not only that Neanderthals—extinct human relatives that either constituted a closely allied species or were a subspecies of *Homo sapiens*—and humans coexisted for thousands of years, but also that they interbred. (Except for those native to Africa, every human on Earth carries a small amount of Neanderthal DNA in his or her genome.) Similarly, when a new fossil is found, scientists are able both to date it with accuracy and to obtain a deeper understanding of its relationship to earlier ones.

Whether we look backward or forward, our understanding of the course of life on Earth remains ever in flux. But Wallace and Darwin's remarkable explanation of the mechanisms of evolution remains at the heart of every discovery.

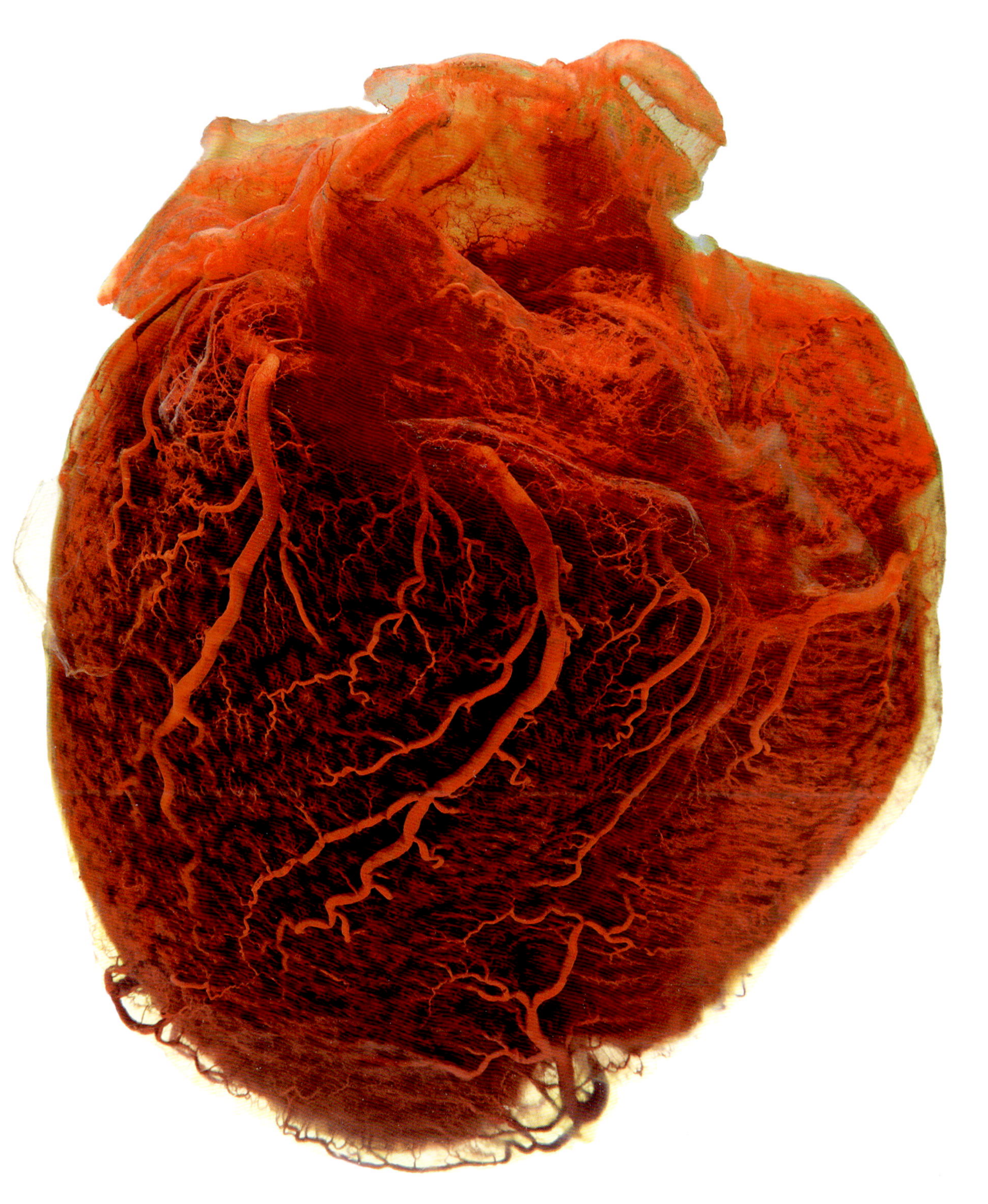

BELOW Even after decades of research, the human brain — and its evolution — remains a subject of intense interest and study. Although the brains of early human ancestors have not been preserved in the fossil record (unlike soft tissue in some other species), the fossil skulls and (in rare cases) natural casts of skull interiors have allowed scientists to estimate that the human brain has tripled in size over seven million years of evolution, with much of that growth taking place in the genus *Homo* over the past two million years. For example, fossils of *Homo habilis* from 1.9 million years ago show an increase in size in a region corresponding to a part of the frontal lobe called Broca's area, which is affiliated with the development of language.

OPPOSITE A stained section of a modern human brain. The increase in brain size over time can be measured through looking at braincase volume. Analyses show that the skulls of early hominids such as *Australopithecus afarensis* had a volume of about 400–550 ml and *Homo habilis* about 600 ml, while modern gorillas' skull volumes are larger, up to 750 ml. From early in its history, however, the skull volume of *Homo sapiens* — modern humans — has measured 1,200 ml or greater, with especially notable increases in the size of brain regions associated with such advanced cognitive functions as speech and problem-solving.

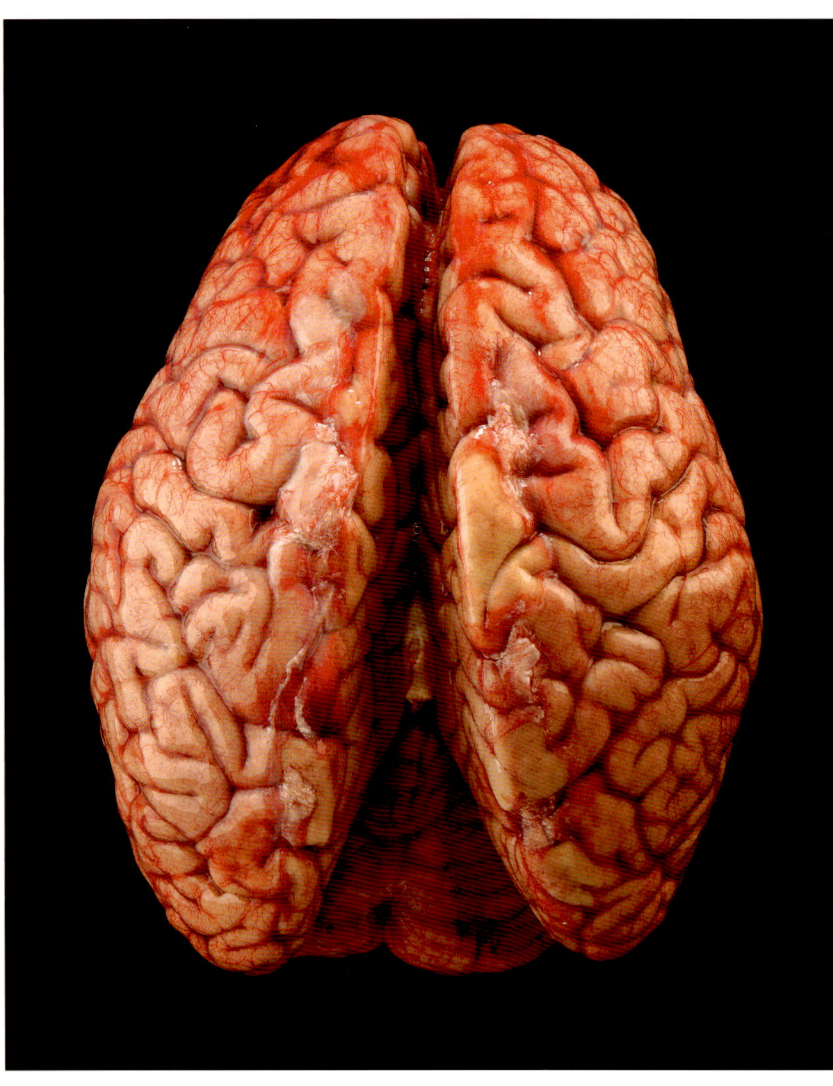

OPPOSITE Nothing has revolutionized the classification of species more profoundly than the study of their DNA. The mapping of entire genomes (an organism's complete set of genetic information) — which has been completed in hundreds of species ranging from sponges and sea anemones to birds and mammals — has in some cases clarified our understanding of species relationships, and in others thrown it into doubt and confusion. Analyzing the genetic makeup of Darwin's finches, which were key in the development of the theory of evolution, for example, has brought to light that the famous birds are not actually true finches — and, on top of that, some may not even form different species. The original model created by James Watson and Francis Crick illustrates the double-helix structure of DNA.

PAGES 200–1 Since its discovery in 2013, a hidden chamber in the Rising Star Cave (part of the Cradle of Humankind World Heritage Site in South Africa) has yielded more than fifteen hundred bones of *Homo naledi*, a previously unknown fossil hominid. The chamber containing the bones — which were in an extraordinary state of preservation — could be reached only through a seven-inch-wide "chimney." As a result, the scientists hired for the Rising Star Expedition (six women dubbed the Underground Astronauts) were selected partly for their build.

PAGE 202 A skull of the early human ancestor *Australopithecus sediba* (about two million years old) from South Africa. The first specimen from the genus *Australopithecus*, named *A. africanus*, was discovered in the same region in 1924, and the decades since have seen the unearthing of fossils of several other related species. Together they paint a portrait of a group (first evolving about four million years ago) that demonstrates significant hominid evolution and that led to the appearance of *Homo,* our own genus.

PAGE 203 A small fraction of the find of *Homo naledi* remains from the Rising Star Cave in South Africa. Unlike many fossil finds, these are so abundant and well preserved (the fifteen hundred–plus specimens are enough to re-create at least fifteen skeletons) that they will provide paleoanthropologists and other scientists with enough research material to occupy them for years. "Judging by the features of the fossil bones," says Lee Berger, who headed the excavation and then described and named the species, "*H. naledi* may indeed be one of the earliest members of our genus."

PAGES 204–5 Bones of *Homo naledi* from the Rising Star trove. The presence of so many fossils of the species so far from the cave's entrance inspires the question: How did they get there? No other fossils of ancient mammals are found at the site, nor are there any signs that the remains were carried into the cave by water. (In fact, many of the small bones are still roughly in their natural positions on the body.) One theory is that *H. naledi* practiced some form of funerary behavior, and the individuals were brought to the cave either soon before or just after death.

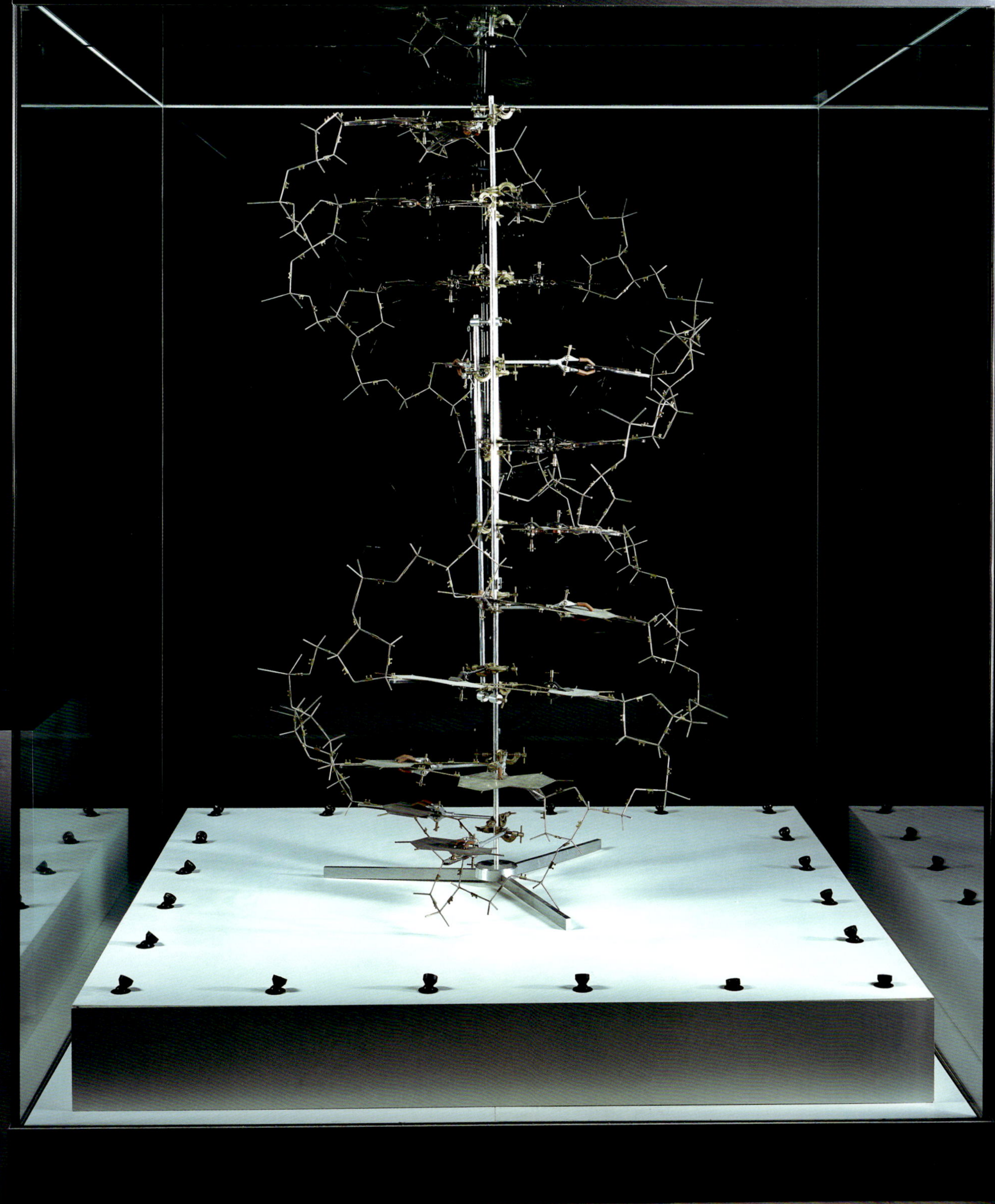

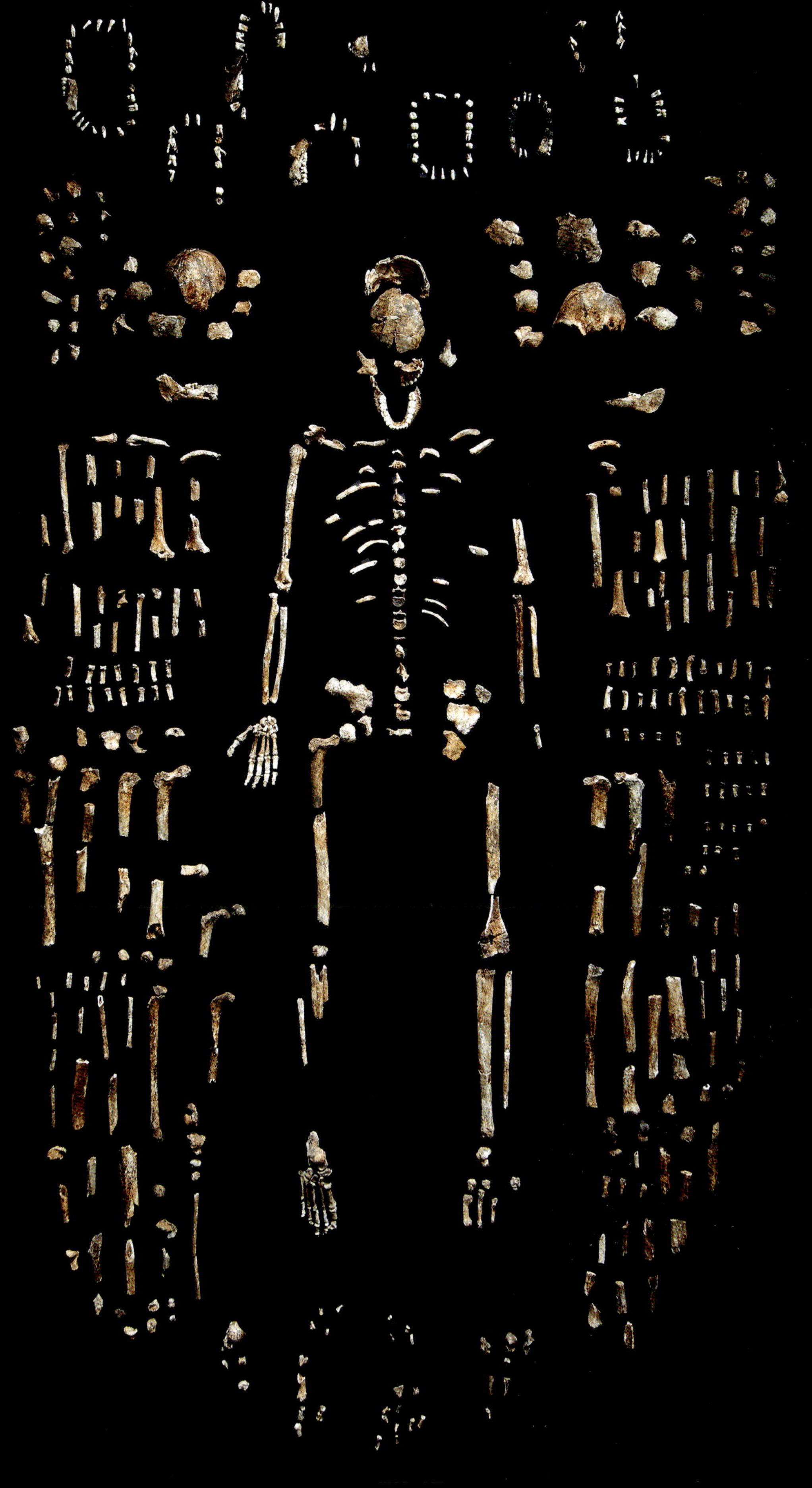

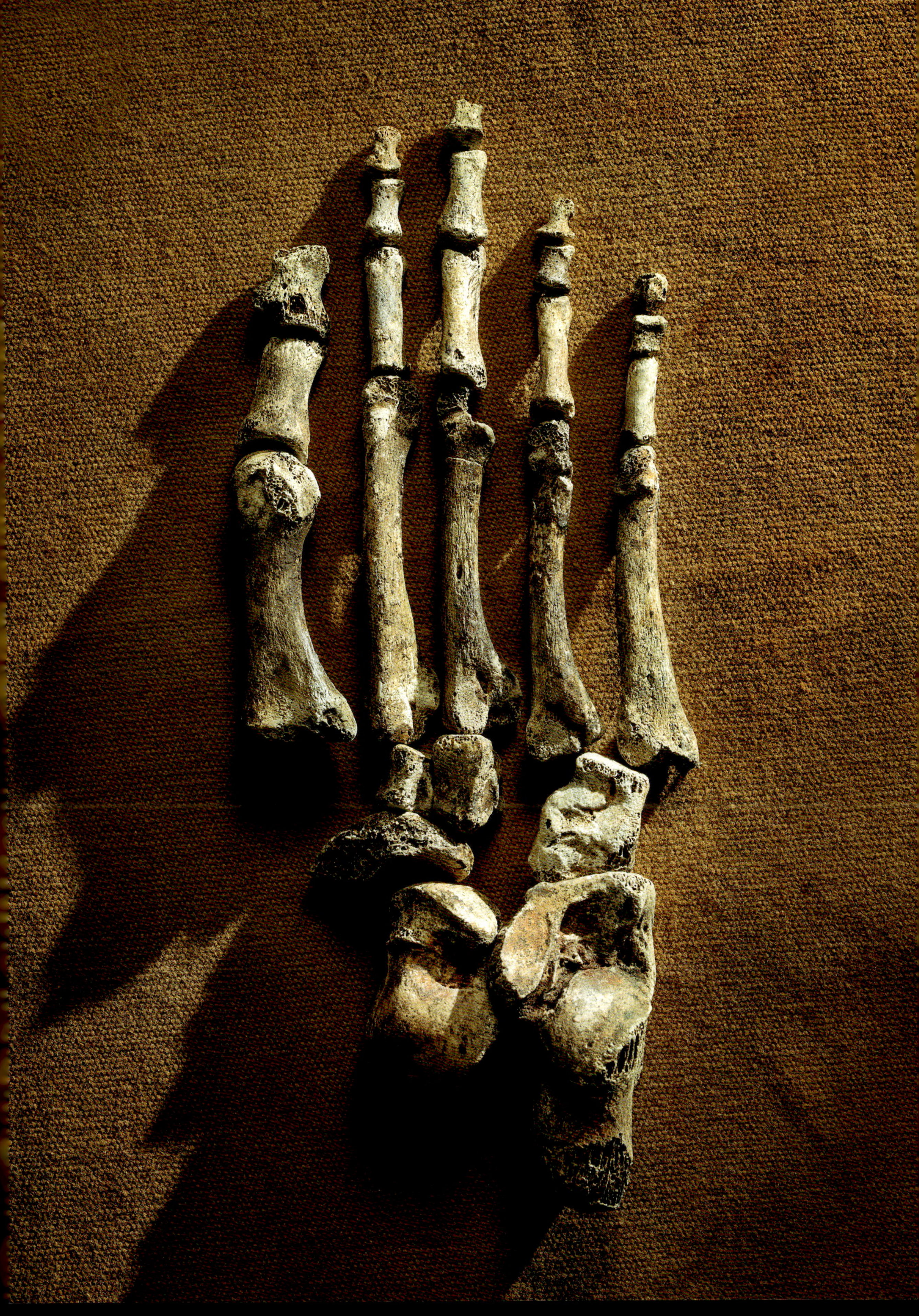

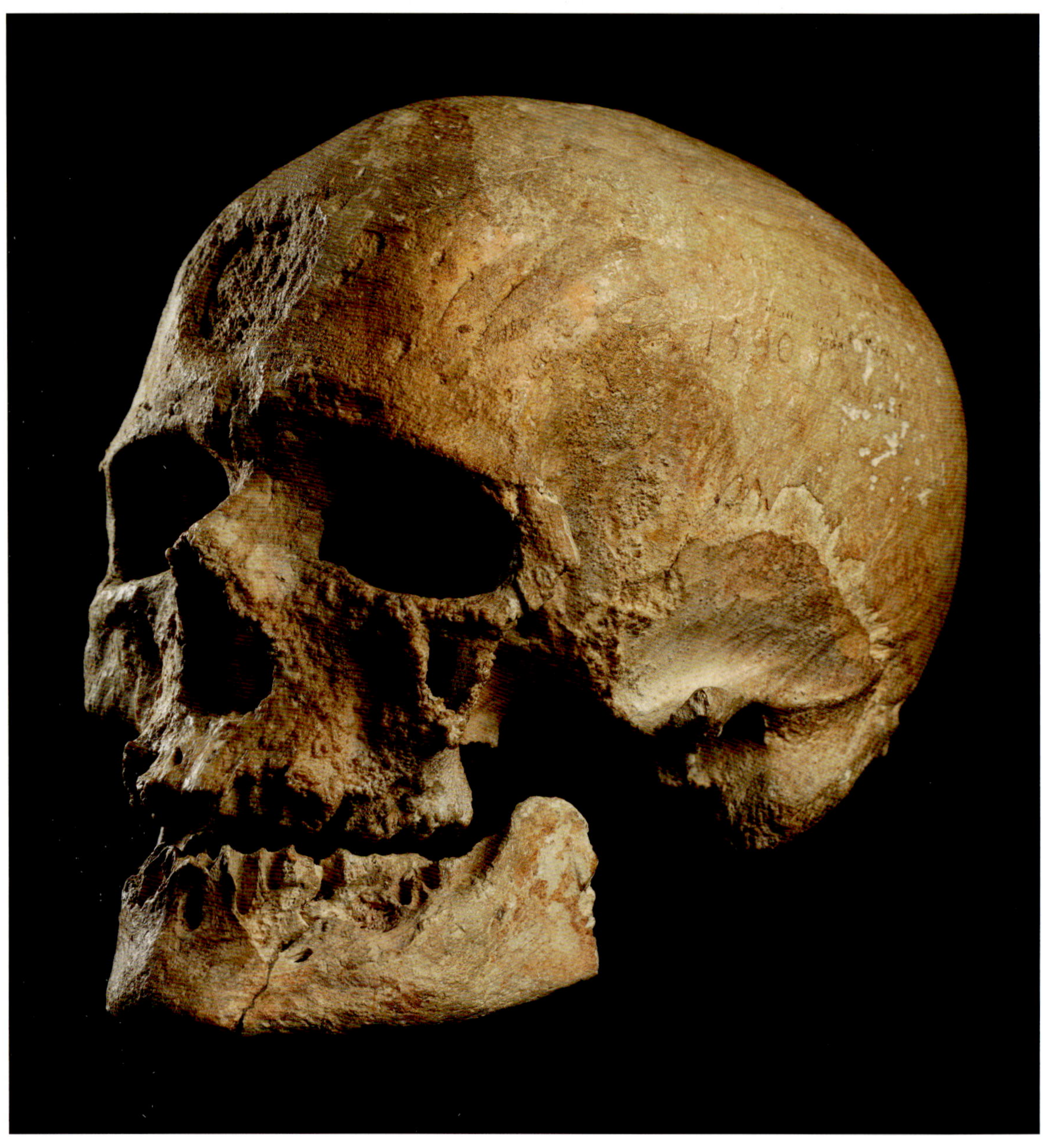

ABOVE Some specimens of ancient *Homo sapiens* (including those represented by this skull) were originally dubbed "Cro-Magnon" after the rock shelter in France where skeletons were found in 1868. Today, with further specimens having been discovered in Italy, Romania, and elsewhere in Europe, they are generally referred to as Early Modern Humans (EMH). There is evidence that they used stone for weapons, ornamented themselves with pigments and shells, and most likely created the stunning cave paintings that are among the most famous and fragile of all ancient finds.

OPPOSITE A Neanderthal foot, which appears very much like our own. Humans and Neanderthals share more than 99 percent of their DNA.

The skulls of two closely related, but different species
of *Homo: H. neanderthalensis* (above), which went
extinct about forty thousand years ago, and the
one extant species in the genus, *H. sapiens* (below).
No one can know for sure why the Neanderthals
went extinct, but theories include climate change
that affected them more drastically than it did early
modern humans; absorption into *H. sapiens* through
interbreeding; and human-caused disappearance,
either gradually, through being out-competed for
resources, or quickly, by means of violent conflict.

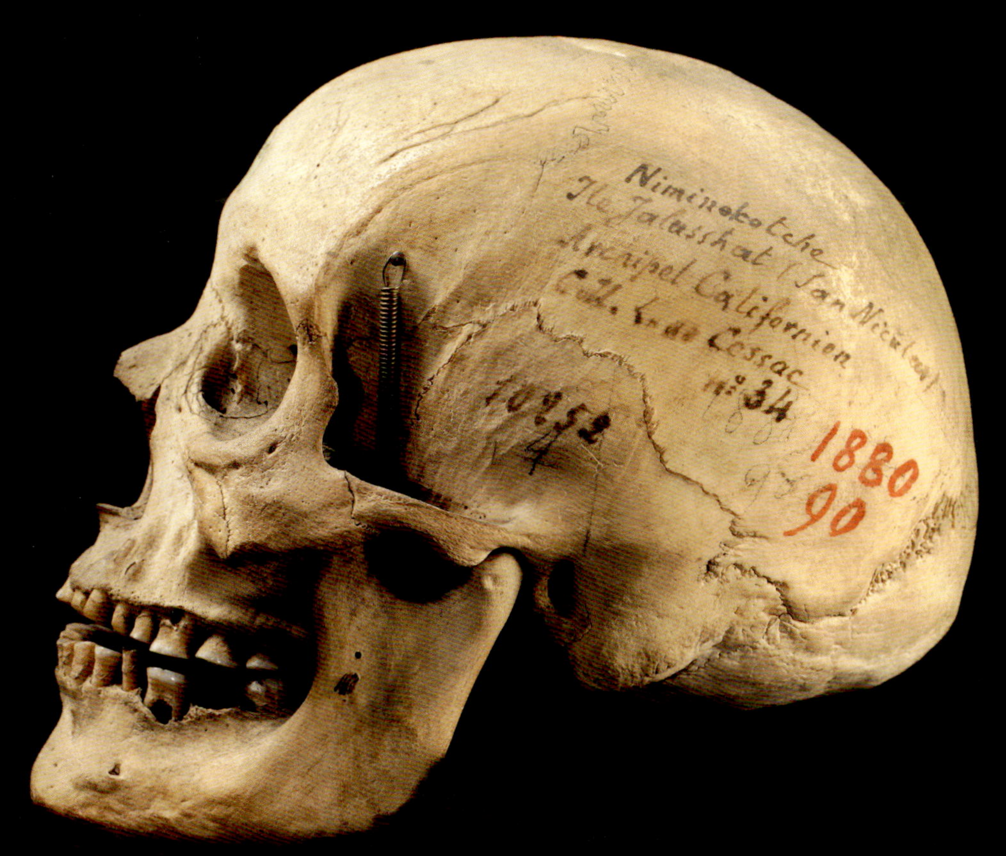

Niminokotche
Ile Jalasshat (San Nicolas)
Arcipel Californien
C.L. de Cessac
Nº 34
1880
90

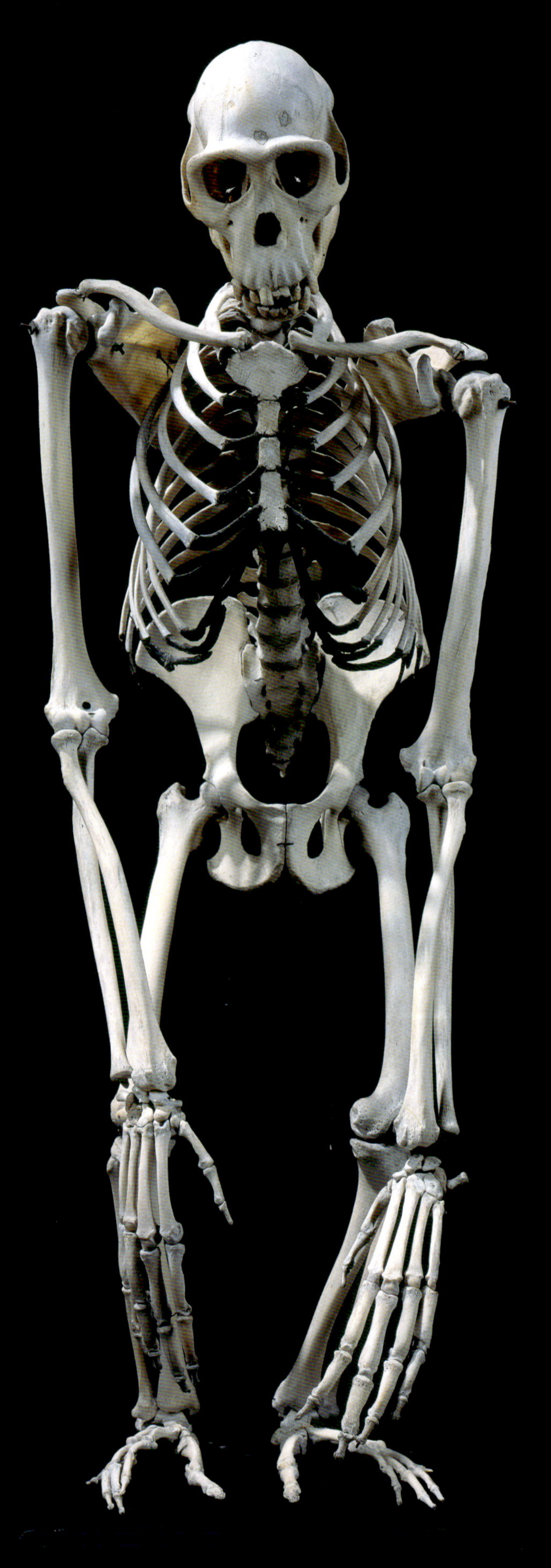

OPPOSITE The skeleton of an orangutan (genus *Pongo)* illustrates both the similarities and the differences between the skeletal structure of humans and the other great apes. The orangutan's squat legs and long arms, fingers, and toes are designed for ease of movement in its preferred arboreal habitat. (The species can both move quadrupedally through the branches and straighten to a crouch to walk along thick branches.) Like humans, orangutans have fully opposable thumbs, but they also have opposable big toes, useful for grasping.

ABOVE AND PAGES 212–13 Among the most evocative of early human remains are the footprints left behind. We can only guess at the purposes and goals of our ancestors who walked past this volcano (at what is now the Engare Sero site near the famous Ngorongoro Crater) and into history. What the 350 preserved footprints do tell us is that two groups walked through wet volcanic ash here about 120,000 years ago. One, heading westward, was made up of eighteen individuals — men, women, and children — all traveling together. The other, heading eastward, was comprised of individuals who may have been traveling separately, walking or running toward a destination that will always remain unknown.

U.S.A.
Leakey Foundation
www.leakeyfoundation
E9

Over the course of millions of years, as the human species has evolved, so have the pathogens that afflict us. These two individuals (Russian prisoners) suffer from multi-drug-resistant tuberculosis, a disease caused by *Mycobacterium tuberculosis*. Specialists in infectious diseases have traced the lineage of tuberculosis as far back as forty thousand years, suggesting that it affected humans at the same time that *Homo sapiens* were spreading from Africa to the other continents. But the disease's evolution has contemporary real-world consequences: Like many other bacteria, over the course of just a few decades *M. tuberculosis* has evolved the ability to resist antibiotics that had previously been effective in eradicating it — which means that scientists must develop new medications, which leads to the bacterium's further adaptation. It is an evolutionary arms race that shows no sign of slackening.

Evidence of natural selection, as seen in a group of young *Homo sapiens:* Dark skin, eye, and hair color were likely selected for in the hot, sunny conditions of East Africa, the birthplace of our species. Once humans spread to other (colder, cloudier) regions, those features were no longer so favored, and other tones began to survive. Some scientists speculate that "personal selection" (e.g., individuals with blue eyes and blond hair being attracted to others with like features) may also have contributed to the diversity in hair, skin, and eyes we see today.

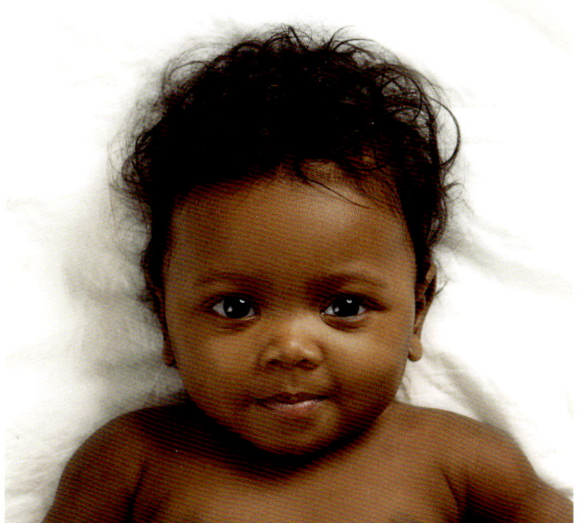

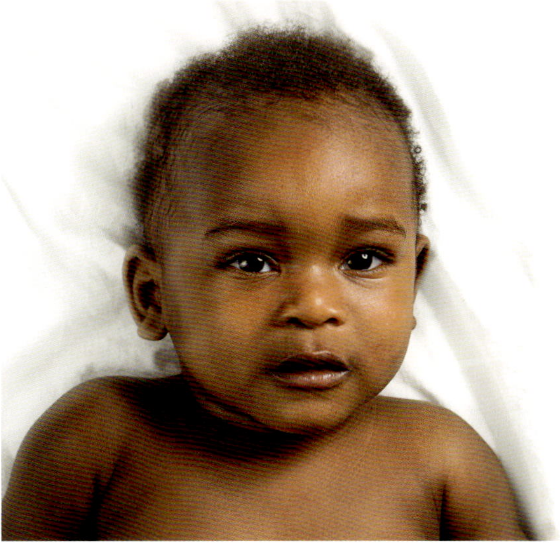

OPPOSITE Genes are always mutating randomly, and the vast majority of these small changes have no effect on the survival of the individual or the species. Most effects are not even noticeable. But albinism is a well-known inherited condition now known to be caused by a variety of mutations in different genes that help control the production of melanin in the skin, hair, and/or eyes. There is as yet no medical means of preventing albinism, but most people with the condition live healthy lives with careful management.

BELOW The possibility that natural selection could produce an organ of such exquisite complexity as the human eye bedeviled even Charles Darwin himself, giving him, as he once wrote to a friend, "a cold shudder." The challenge: In his time, the intermediate evolutionary steps that might lead to the human eye were unknown, or at least not understood. But Darwin never wavered in his opinion that eyes — like every-thing else on Earth — had, in fact, evolved through the process whose theory he had explained. And today, on molecular and other levels, scientists are untangling the complexities of eyes (human and otherwise), proving that, once again, Darwin was right to hold to his beliefs.

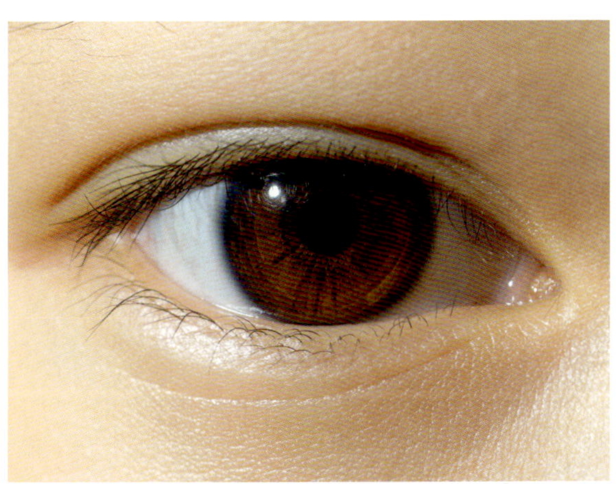

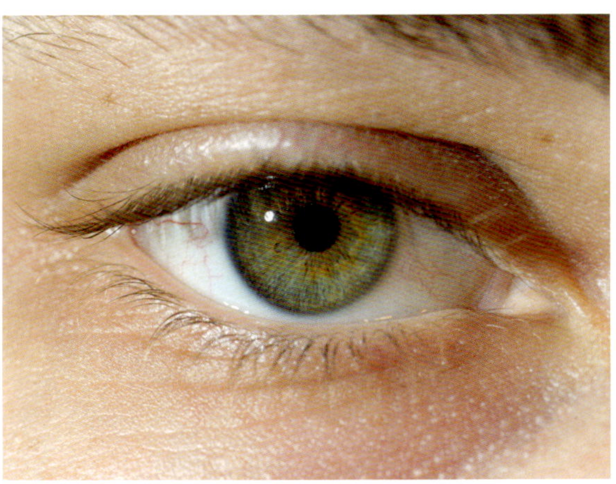
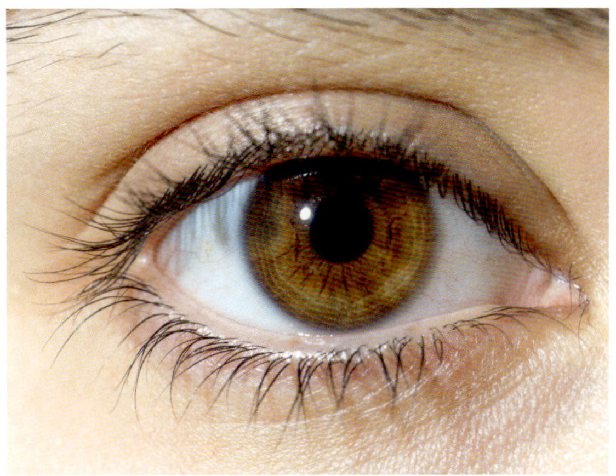

THE EVOLUTION OF EVOLUTION

Sometimes it feels as if the study of evolution is perpetually looking backward. This is easy to understand: The fossil record, as incomplete as it is, provides a fascinating account of what used to exist; a window on how life on Earth blossomed and evolved through mutation and natural selection; and, strikingly, a view of the mass extinctions that have periodically swept the planet.

Such extinction events, followed by renewed bursts of diversity as new species filled emptied ecological niches, have played an enormous role in shaping the history of life. For example, the end of the Permian Period saw the disappearance of more than 90 percent of all species, while countless others, including the non-avian dinosaurs, disappeared at the end of the Cretaceous Period. (This most recent extinction event was followed by the appearance of the earliest ancestors of modern dogs, cats, whales, camels, rhinos, and many others—including humans.)

Today, by studying evolution in real time, as it occurs, scientists hope to gain insights into a possible new mass extinction, the sixth in geologic history, which many suggest is occurring now. A deeper understanding may allow us to forestall some of its effects—or help us learn what is likely to come next, and whether our species will be a witness to—or a casualty of—what is taking place.

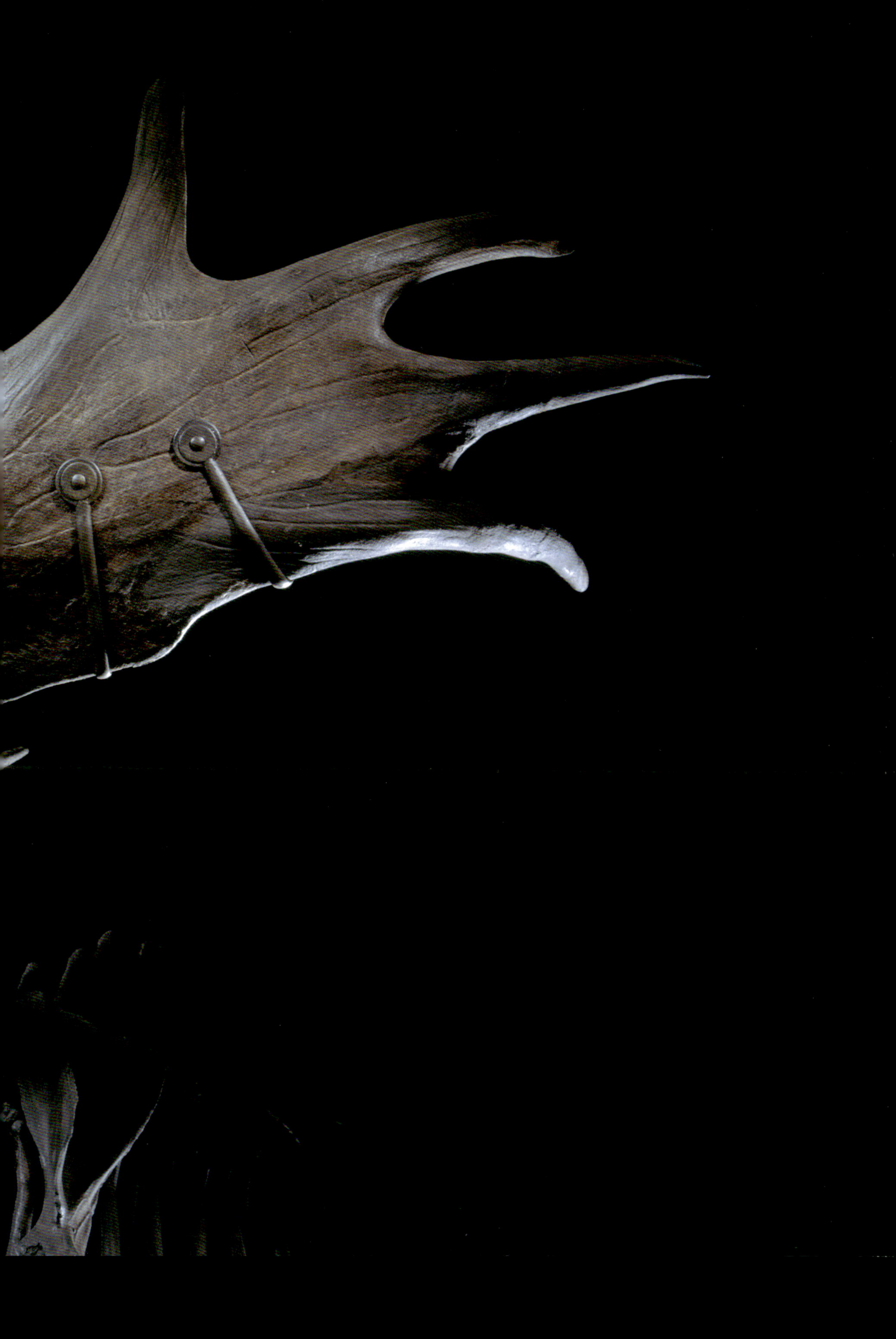

PAGE 220 The process of evolution is an endlessly unfolding marvel that invites both close study and wide-eyed wonder. Just as it has after each of the great extinctions that have periodically struck the planet's flora and fauna, evolution, barring any planet-wide catastrophe, will continue far into the unimaginable future. But will such vulnerable species as this red-eared guenon *(Cercopithecus erythrotis)* — or humans *(Homo sapiens)* — be there to witness it?

PAGES 222–23 Beginning in 1909, scientists and preparators took on the enormous task of creating this great elephant display and the other vivid dioramas in the American Museum of Natural History's Hall of African Mammals. Even then, some were already warning that elephants — and many other species found on the continent — were under grave threat of extinction. Today, those threats have only grown, as the trade in ivory and bushmeat, human population growth, and habitat destruction have reduced populations of elephants and other wildlife to a fraction of what they were a century ago.

PAGES 224–25 The magnificent — and extinct — Irish elk *(Megaloceros giganteus)*. So named because some of the most famous specimens have been found in peat bogs in Ireland, this gigantic deer was found throughout Europe, parts of Asia, and northern Africa during the Pleistocene Epoch. Since it was so huge and lived in Europe, the Irish elk was a center of the eighteenth-century scientific controversy over whether any species ever went extinct. Thomas Molyneux, who named the animal, voted no: "That no real species of living creatures is so utterly extinct, as to be lost entirely out of the World, since it was first created, is the opinion of many naturalists." But Charles Darwin and others wondered: If the Irish elk actually still survived, where was it? (A living seven-foot-tall deer with antlers spanning twelve feet wouldn't have been easy to hide.)

OPPOSITE When the Pilgrims settled in New England, they found an abundant population of the heath hen *(Tympanuchus cupido cupido)*, a large game bird now known to be a subspecies of the prairie chicken of the Midwest and West. In the late eighteenth century, heath hen populations earned them the dismissive name "poor man's food," with household servants supposedly negotiating *not* to be served the birds more than two or three times a week. As has happened innumerable times with creatures that humans consider tasty, however, by the mid-nineteenth century the "poor man's food" had become rare. The last population of the heath hen, confined to the island of Martha's Vineyard, hung on for a few more decades, with the last member of the species, Booming Ben, dying in 1932.

PAGES 228–29 The great auk *(Pinguinus impennis)* is a shining example of two phenomena seen around the world: 1) Islands are home to many of the planet's strangest and most extreme plant and animal species, and 2) Those species have gone — and are going — extinct at an astonishing rate. The great auk was a 2.5-foot-tall, flightless, unwary, penguin-like relative of the puffin that lived on islands in the North Atlantic. A human food source for tens of thousands of years — cleaned auk bones have been found in Neanderthal middens — the species survived until people decided that auk down made great filling for pillows. (Sailors could simply pick up the tame birds and pluck them.) Despite attempts to protect the species, the auk was extinct by around 1850.

LEFT Few stories of extinction are more spectacular than that of the passenger pigeon. Once thought to be the most abundant bird in North America, with a population that could reach five billion, the species' flocks famously could darken the skies for hours at a time as they flew overhead. But the pigeons were hunted relentlessly for food and sport, especially on their nesting grounds, throughout the nineteenth century. Combined with deforestation and other factors, this onslaught led with shocking speed to the species' extirpation. After the death of Martha, the last surviving captive pigeon, in 1914, all that was left of those astonishing flocks were museum specimens such as this.

OPPOSITE By now, it's widely known that beetles range across the earth in both great abundance and great diversity. The question is why? Why *are* there so many different kinds? Scientists have long theorized that the main cause of their success was a tendency for new beetle species to form. Recently, however, researchers studying the fossil record have found that beetle families appear to have suffered far lower family-level extinction rates over time than have many other animals. In fact, not a single family in the gigantic suborder Polyphaga (which includes everything from scarab beetles to ladybugs) appears to have ever died out, not even during the great extinction event that brought an end to the non-avian dinosaurs. It seems that beetles are actually extinction-resistant.

INDEX

PHOTOGRAPHER'S BIOGRAPHY

Robert Clark is an award-winning photographer whose magazine work includes more than forty stories for *National Geographic*. Clark's work also appears in several notable books, including *Friday Night Lights, First Down Houston,* and *Feathers: Displays of Brilliant Plumage*. His photographs documenting the September 11, 2001 terrorist attacks are some of the most widely reproduced images of the event and are part of the collections of many museums and archives. Clark lives in Brooklyn, New York, with his wife and daughter. Learn more about his work at robertclark.com.

ACKNOWLEDGMENTS

This book would not have been possible without the dozens of assignments I have been given by the *National Geographic*. I was fortunate to work on the story "Was Darwin Wrong?" (Nov. 2004), which sparked my interest. Other stories came my way and fed my interest in evolutionary biology.

Working on the rise of mammals, dinosaur behavior, the human brain, the human heart, early human finds, and the evolution of feathers, of whales, and of dogs — and so many other stories — helped to create a through-line in my work that provides the bulk of this book.

The list of people who inspired me and helped on these stories is long yet crucial to any insight that might come from this work.

I'd like to acknowledge Chris Johns, Bethany Powell, Kurt Mutchler, Kathy Moran, David Griffon, Sarah Leen, Bill Marr, Alice Gabriner, Kim Hubbard, Susan Welchman, Laura Lakeway, Bill Allen, Jamie Shreeve, and writer Carl Zimmer at the Geographic.

I have enjoyed working with David Quammen, whose foreword for this book shows his ability to cut to the point in the most descriptive and beautiful words, making it sound easy. I have learned from his books and articles, which are fluid and inspiring.

Thanks to Phaidon, especially Deb Aaronson — for giving this book life and for her keen eye — and Bridget McCarthy, who kept it on the tracks. Sharon AvRutick gave shape to my ideas, and Joseph Wallace wrote the words that help explain the visuals I provided. Miko McGinty and Rita Jules created this beautiful design.

I am grateful to the countless scientists, researchers, museum technicians, and public information officers I've worked with: Reese Barrick, Laura Wilson, and Curtis Schmidt at the Sternberg Museum in my hometown of Hays, Kansas; George Beccaloni of the British Museum for his special insight into Alfred Russel Wallace; Carla Dove at the Smithsonian Institution; Dr. Michael Novacek, Roberto Lebron, and Will Harcourt-Smith at the American Museum of Natural History; Gerald Mayr at the Senckenberg Research Institute and Natural History Museum; Dan Gordon and Victoria Page at the Great North Museum: Hancock; and Lee R. Berger at the University of the Witwatersrand (Johannesburg).

Special thanks to Carl Mehling from the AMNH for his careful reading.

Thank you is not enough to my parents, Russ and Dora Lou Clark, and my sisters and brothers, Lynn, Cindy, Steve, Sara, and Pat. My wife, Lai Ling, and daughter, Lola, make every day an adventure worth having. Michael, Marlene, Ed, Dan, Eva, and Luca listened to my ideas, both good and bad.

Thank you to Parker Feierbach, AJ Wilhelm, Christopher Farber, Alex di Suvero, Jill Lewis, and David Coventry, who diligently assisted me.

A special note about the influence that Todd James at the Geographic has had: We have worked together on several stories but his passion for "Was Darwin Wrong?" was contagious and has informed other portions of this work.

And to the readers of this book: Pass along this information. I truly believe it is the greatest story ever told.

—Robert Clark

FOR LOLA

Phaidon Press Limited
Regent's Wharf
All Saints Street
London N1 9PA

Phaidon Press Inc.
65 Bleecker Street
New York, NY 10012

phaidon.com

First published 2016
© 2016 Phaidon Press Limited

ISBN 978 0 7148 7118 9

A CIP catalog record for this book is available from
the Library of Congress and the British Library.

Editor: Sharon AvRutick
Editorial Coordinator: Bridget McCarthy
Production Controllers: Sue Medlicott,
    Nerissa Dominguez Vales, Matt Harvey
Design: Miko McGinty and Rita Jules

Printed in China

SPECIAL THANKS
The photographer is grateful to the following
institutions for granting access to photograph their
collections: American Museum of Natural History
(New York); British Museum (London); Butterfly House
(Missouri Botanical Gardens, St. Louis); Cincinnati Zoo
& Botanical Garden; Cleveland Metroparks Zoo; Down
House (Kent); Fort Wayne Children's Zoo; Museum
of Natural History (Paris); Natural History Museum
(London); Natural History Museum at Tring; Oxford
University Museum of Natural History; Royal Botanic
Gardens, Kew (London); Ruth Hall Museum of Paleon-
tology (Ghost Ranch, New Mexico); Science Museum
(London); Smithsonian Institution; Sternberg Museum
of Natural History (Hays, Kansas); University of
Michigan Museum of Natural History (Ann Arbor);
William L. Clements Library, University of Michigan
(Ann Arbor).

CAPTIONS
COVER The moth *Xanthopan morganii* and Darwin's
orchid *(Angraecum sesquipedale),* which it pollinates.
(See page 127.)

PAGE 2 Each new detail learned about an extinct
species adds one more piece to the vast mosaic of
our understanding of life on Earth and how it evolved
over time. For example, dinosaur trackways (like
these found near Villa El Chocón, Argentina) allow us
a glimpse of long-vanished animals' habits, bringing
the distant past closer for at least a moment.

PAGE 4 The skeleton of a prehistoric whale demonstrates that it was descended from four-legged land
mammals: The fins evolved from front limbs, while the
hindlimbs — though still visible — have become tiny.